Quazilinearization and the Identification Problem

Quasilinearization and the Identification Problem

Richard Bellman
Professor of Mathematics,
Electrical Engineering and Medicine
University of Southern California

Robert Roth
The Charles Draper Laboratory, Inc.
Cambridge, Massachusetts, 02139

World Scientific

World Scientific Publishing Co Pte Ltd
P O Box 128
Farrer Road
Singapore 9128

9971-950-44-8
9971-950-45-6 pbk

Printed in Singapore by Kim Hup Lee Printing Co. Pte. Ltd.

To Our Wives

Nina and Micheline

PREFACE

The purpose of this book is to present an overview of the techniques of quasilinearization as they are applied to problems of system identification. While the ideas of quasilinearization were developed as a numerical tool for solving problems defined by nonlinear differential equations, these same ideas exhibit great potential for solving identification problems. As we intend to show in this book, the quasilinear technique, when viewed as an identification procedure, has inherent advantages in establishing the intricate interrelationships which exist in complex physical systems.

Quasilinearization, by its very construction, allows us to compute the solution to a system of differential equations by an iterative procedure which roughly parallels the Newton-Raphson technique for finding the roots of a nonlinear continuous function $f(x) = 0$.

In the early 1960's the relation between quasilinearization and the identification problem was studied by Bellman within the context of differential equations. By observing that if the parameters of a system of differential equations are unknown, then we are quickly confronted with an identification problem. The problem becomes well posed when the output of the system is known either by observation or by design. Within the scope of identification, we are asked to determine those parameters whose solution fits the observations in some optimal sense.

This book is a culmination of our thoughts on identification over the past twenty years.

The text is loosely organized into three areas: First, we consider the problems inherent in the formulation of mathematical models of systems by differential equations and the essential features of quasilinearization. We then consider several non-trivial samples to illustrate the ideas. Finally, we discuss several advanced topics which are central to the quasilinear technique.

Throughout the text we emphasize that it is essential that the mathematician understands something of the underlying process. This understanding furnishes valuable clues to the analytical and computational solution and often yields an excellent first approximation.

Chapters I and II develop the methods and techniques which we use in both modelling the system under study and the quasilinearization techniques we intend to employ throughout the book. Chapter III applies the technique to the problem of orbit determination. Quasilinearization, in its own right, is a very valuable tool for solving difficult problems occurring in the realm of differential equations. Of particular interest is the N-point boundary value problem where the solution is known at N points along the trajectory and we are required to find the solution satisfying this condition. Orbit determination is a classical example of this problem.

Chapter IV presents a simple but direct application of quasilinearization to identification. In a known model of a chemical reaction, we wish to determine the rate constants producing the best theoretical results as compared to a set of observations.

Often, after observing the behavior of an interesting physical phenomenon, we are interested in fitting a mathematical model to the observations. The problem is what model do we choose? Chapter V is concerned with this problem as we examine the intricate complications surrounding the proper modelling of problems in ecology.

Chapter VI illustrates the quasilinear technique in studying the large complex system defining the dynamics of induced enzyme synthesis. Here we wish to determine both initial conditions and system parameters. Our interest in this chapter is to review the complex biomechanics involved and solve the identification problem by quasilinearization.

Chapter VII deals with the identification of distributed systems described by partial differential equations. The existence of more than one space dimension greatly complicates the solution of such a problem. Clearly new methods must be applied to make these problems trackable.

In Chapter VIII we return to the problem of what is a good mathematical model for a given set of observations. These ideas are applied to Chapter IX with the measurements of heart parameters.

A second technique, besides quasilinearization, which is very useful in the identification of systems is dynamic programming which is considered in Chapter X by studying the inverse problem in Transport Theory. The use of dynamic programming in this instance not only clearly defines a computing algorithm for solving the transport problem, but allows us to define the problem in a new light using invariant embedding.

A short comment, comprising Chapter XI, defines the inverse problem for several settings in mathematical physics.

In Chapter XII, we return to the consideration of theoretical developments and, in particular, the techniques for handling partial information. At this point in the book, we feel enough work has been established to allow us to consider a very important part of the identification problem, structural identifiability, as discussed in Chapter XIII. Chapters XIV and XV consider both the initial approximation and the criteria for the data fit.

There is a wide body of problems which dominate the areas of mathematical biology, linear dynamics and systems with built-in delays in which the basic solution can be written as a sum of exponentials. We devote Chapter XVI to an examination of this problem.

We end the book with two additional examples. Chapter XVII is a prototype of the problem of finding a fault in a complex system, that is finding two defective coins. Chapter XVIII illustrates a difficult engineering problem, that of estimating a dynamic coefficient of friction under actual test conditions.

Our aim throughout this book has been twofold. First, to

demonstrate the power of the quasilinear technique when it is applied to identification problems and second, to show, by example, the wide range and complexity of problems which can be solved. It is clear from reading the text that much remains to be done. There is much opportunity for research in these areas.

Richard E. Bellman
Santa Monica, Calif.

Robert S. Roth
Cambridge, Mass.

CONTENTS

CHAPTER XV: DETERMINING A CRITERION

CHAPTER XVI: SUMS OF EXPONENTIALS

CHAPTER XVII: FINDING TWO DEFECTIVE COINS

CHAPTER XVIII: THE MOTION OF A ROCKET IN A SMOOTHBORE LAUNCHER

Chapter I

MATHEMATICAL MODELLING

1. Introduction

In this chapter we shall discuss a variety of problems which are associated with the procedures of developing mathematical models of processes which have their common origins in functionally realizable systems.

We shall, as far as possible, examine the many facets of analysis which are required in order to develop a mathematical model of a physical system. Furthermore, we wish to make precise the concepts inherent in the theories of system identification with which we will be concerned throughout this book.

Before a systematic study of mathematical models is attempted, we would like to dwell on the basic ideas of systems as they occur in the real world. Fundamentally we prefer to define a system as a set of elements plus a collection of rules governing the interrelations between the elements, and the behavior of the elements over time. With this broad definition we can begin our investigation of the properties and behavior of such systems and in particular, we shall be dealing with methods of identifying such systems when direct observations of their behavior can be made.

We can identify a system by its mathematical structure. In making this identification, we hope that the actual system can be approximated by a mathematical system which can be precisely defined.

2. Mathematical Structure

Let us begin with a description of the mathematical model of a typical physical system which can be observed in some manner. We suppose that the physical system under observation varies over time in such a way that all the information that we ever wish to have concerning the system is contained in a set of N functions of time, $x_1(t)$, $x_2(t),\ldots,x_N(t)$.

The quantities $x_i(t)$, $i = 1,2,\ldots,N$ will be called the state variables of the system and N its dimension. It is occasionally useful to regard the x_i's as the components of a vector X, the state vector. We write it as a column vector.

$$X = \begin{pmatrix} x_1 \\ x_2 \\ \cdot \\ \cdot \\ \cdot \\ x_N \end{pmatrix} . \tag{2.1}$$

We now wish to study the behavior of the state vector as a function of time. More ambitiously, given the present state and the past history of the system, we would like to predict its future behavior. The associated identification problem would be to determine the precise system from which the known time history of the state vector $X(t)$ originated.

Initially we shall choose those systems which are governed by ordinary differential equations. Further we shall not ask for the value $X(t_1)$ at some $t_1 > t$, but rather for the instantaneous change in $X(t)$

as a function of the past and present states. We thus write the equation of state in the form

$$\frac{dX}{dt} = G(X) \quad , \tag{2.2}$$

where $G(X)$ is a functional of both present and past vector states.

In order to simplify the problem, by writing a more manageable equation, we boldly assume that $G(X)$ is independent of the past history of the process and depends only upon the present. In other words we use our defining equation as a differential equation of the form

$$\frac{dX}{dt} = H(X) \quad , \tag{2.3}$$

where $H(X)$ is a vector function of the current state X. This equation is a shorthand notation for N simultaneous equations of the form

$$\frac{dx_i}{dt} = h_i(x_1, x_2, \ldots, x_N) \ , \qquad i = 1,2,\ldots,N \quad . \tag{2.4}$$

To appreciate the magnitude of this assumption that we have made, observe that according to well-known mathematical results concerning the existence and determination of solutions of differential equations, under reasonable assumptions concerning the functions $h_i(x_1, x_2, \ldots, x_N)$, we can assert that a system governed by Eq. (2.4) is uniquely specified at any future time t_1 by its state at some previous time t_0.

In many physical situations this is a manifest absurdity. It is, however, a useful if patent absurdity, and consequently one that we shall base our studies on for a while. The fault in so many mathematical studies of this type lies not so much in sinning as in a lack of realization of the fact that one is sinning, or even in a lack of acknowledgement of any conceivable type of sin.

A mathematical model of a physical process is to be used as long as it yields a prediction in reasonable accordance with the observation.

Considering the complexity of actual physical phenomena vis-a-vis the simplicity of our mathematical models, it is not surprising that we are forced to modify our formulation from time to time in order to obtain more accurate results. What is remarkable is that deep understanding of many physical processes can be obtained from rudimentary assumptions.

3. Governing System Equations

In the last section on mathematical structure, we have limited the scope of mathematical systems to those governed by the vector set of equations,

$$\frac{dx_i}{dt} = h_i(x_1, x_2, \ldots, x_N) , \qquad i = 1,2,\ldots,N \ . \tag{3.1}$$

It is important to note that a great deal of information is contained in mathematical structures such as those given in (3.1) and, indeed, the bulk of our investigations into the identification problem rests squarely on this simple looking structure.

We shall assume throughout the book that the mathematical model will always be continuous and therefore the set of functionals $h_i(x_1(t), x_2(t), \ldots, x_N(t))$ will be continuous not only in the independent time variable t but, as functions h_i, will be continuous in the functions $x_i(t)$, $i = 1,2,\ldots,N$. Furthermore, we shall not require that the system (3.1) be necessarily linear for much of the rich complexity of mathematical models are embedded in the nonlinearities of the system.

A very instructive way to consider the mathematical structure of a complex system is to represent the system diagrammatically as shown in Fig. 1.1.

A diagram such as the one shown in Fig. 1.1 can prove to be essential in studying large systems, such as the analysis of induced enzyme synthesis which is discussed in Chapter IV. In a very clear way, the diagram points out the deep structure of a mathematical system and gives us valuable clues concerning the identification procedures which could be defined for the identification of substructures.

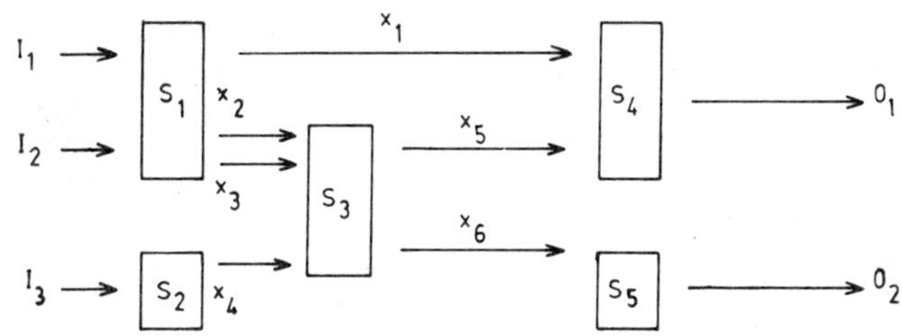

Fig. 1.1 A Block Diagram of a Complex System

By saying that each substructure $S_1, S_2, \ldots$ is made up of a vector differential equation of the form (3.1) we may construct a very rich complex system. A single N-th order ordinary differential equation is itself a very complex structure for it can potentially define N different solutions, each of which has a time dependent character which depends on the parameters of each subsystem.

4. Initial and Boundary Conditions

In the theory of ordinary differential equations of the form,

$$\frac{dx_i}{dt} = h_i(x_1, x_2, \ldots, x_N) \;, \qquad i = 1,2,\ldots,N \;, \tag{4.1}$$

the solution is obtained only after a definite set of initial conditions is defined. In fact the set of equations (4.1) defines a class of solutions and the most common way to extract one particular solution from the class is by assigning a particular set of initial conditions to be associated with the set of equations.

The identification problem requires us to look at the class of solutions to a set of differential equations in a different light. In this situation the initial and/or boundary conditions are not known to us but rather become the very things sought. Now our ability to construct a solution to the required set of differential equations has broadened, but with this sudden freedom arises the associated problem of determining the selection of a proper combination of candidate solutions. It is this problem which will become our main concern in later chapters.

5. The System Parameters

Not only can the system behavior be altered by a change in the initial and boundary conditions, but it can also be altered drastically by the system parameters.

An interesting example, which we will return to later is the behavior of the Van der Pol equation. This equation can be written as

$$\frac{d^2u}{dt^2} + \lambda(u^2 - 1)\frac{du}{dx} + u = 0 \,, \qquad \lambda > 0 \,. \tag{5.1}$$

In this example, when the parameter λ is very small, the response is almost periodic (see Fig. 1.2),

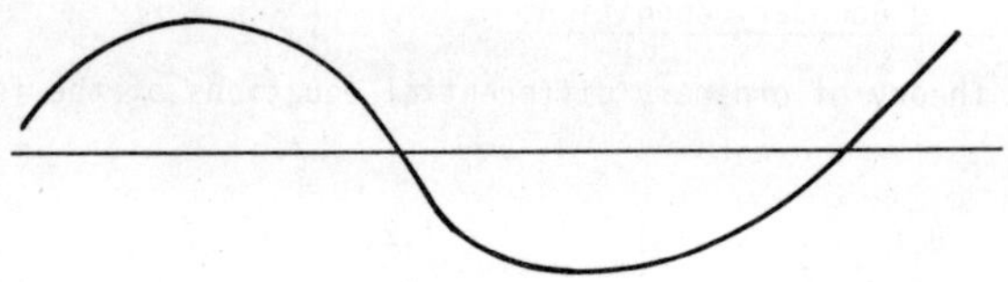

Fig. 1.2 Response of the Van der Pol equation when λ is small

while when the parameter λ is large, the response becomes that shown in Fig. 1.3.

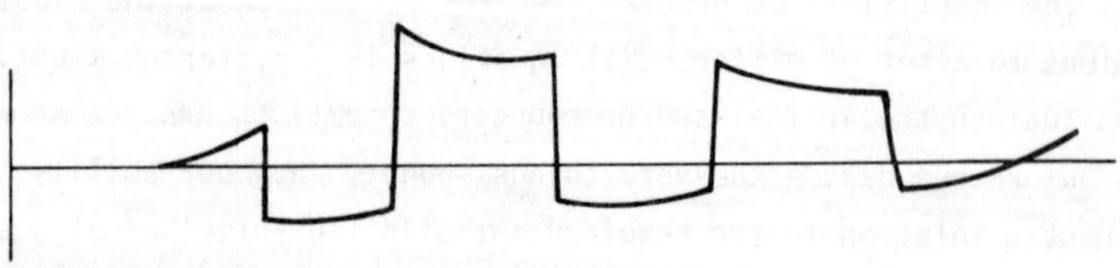

Fig. 1.3 Response of the Van der Pol equation when λ is large

These examples and countless others occurring in engineering problems illustrate strikingly the importance of obtaining a good estimate for the system parameters.

It is probably fair to say that the correct identification of the system parameters, rather than the initial conditions, constitute the bulk of the interest in the identification of systems.

6. The Identification Problem

Consider a vector differential equation

$$\frac{dX}{dt} = G(X,A), \qquad X(0) = C \quad , \tag{6.1}$$

where X is N-dimensional and A is M-dimensional. Suppose that we are given a set of observations r_i, $i = 1,2,\ldots,N$. The identification problem is to determine A and C on this basis.

There are several distinct classes of problems that may be distinguished here. The first is where we suppose that $X(t)$ does indeed satisfy any equation such as (6.1) and that A and C exist. The second is a design problem where A and C are to be determined so that $X(t_1) \cong r_i$, $i = 1,2,\ldots,N$. The third is where we know in advance that $X(t)$ does not satisfy an equation of as simple a type as (6.1), but where we want to find the best approximate equation of this structure.

We determine A and C by the condition that the quantity

$$\sum_{i=1}^{N} (X(t_1) - r_i)^2 \tag{6.2}$$

be minimized. This analytic problem, which can be handled in various ways, is common to all three classes of questions.

A systematic approach to the multi-dimensional minimization problem described in (6.2) is supplied by the theory of quasilinearization, which in a number of cases yields techniques which can be considered to be extensions of the Newton-Raphson-Kantorovich approximation method.

The basic idea in the case of a scalar equation is quite simple. Let $a^{(0)}$, $c^{(0)}$ be initial guesses as to the values of a and c and let $x^{(0)}$ be computed as the solution of

$$\frac{dx^{(0)}}{dt} = g(x^{(0)}, a^{(0)}), \qquad x^{(0)}(0) = c^{(0)} \quad . \tag{6.3}$$

These values are obtained either on the basis of experience and intuition, through the use of simpler models, by way of search techniques, or a combination of all three. The next approximations are obtained in the following fashion: Expand $g(x,a)$ around $x^{(0)}$, $a^{(0)}$, keeping zero and first-order terms, and let $x^{(1)}$ be determined as the solution of

$$\begin{aligned} \frac{dx^{(1)}}{dt} &= \frac{\partial g}{\partial x}(x^{(0)}, a^{(0)})(x^{(1)} - x^{(0)}) \\ &\quad + \frac{\partial g}{\partial a}(x^{(0)}, a^{(0)})(a^{(1)} - a^{(0)}) \quad , \qquad x^{(1)}(0) = c^{(1)} \quad . \end{aligned} \tag{6.4}$$

The scalars $a^{(1)}$ and $c^{(1)}$ are now obtained by means of the minimization of the quadratic form

$$\sum_{i=1}^{N} (x^{(1)} - r_i)^2 \quad . \tag{6.5}$$

We have thus reduced the original optimization problem to a succession of operations involving the numerical solution of linear differential equations and the numerical solution of linear algebraic equations. These operations can be carried out accurately and quickly for systems of quite high dimensions. A detailed discussion of this method, with numerous applications, will be found in later chapters.

7. Concluding Remarks

Let us consider briefly some measurement and data processing aspects of large systems. Suppose we are attempting to implement a measurement strategy at time t which depends upon a knowledge of the components of the state vector $x(t)$. If the dimension of x is large, we must accept the fact that an appreciable amount of time will be involved in the measurement of the required information for the identification. It follows then that there are two extremes: act instantaneously on the basis of minimum information; and act on the basis of maximum information at a later time. Both extreme policies will introduce errors and associated costs. Various simple models of control and decision processes of this type will yield analytic estimates which are similar in structure to the classical uncertainty principles of quantum mechanics.

The multi-stage nature of the identification and control process enables us to diminish considerably the effects mentioned above. One might expect that similar multi-stage measurement techniques applied in physics would reduce the uncertainties in quantum mechanics. In any case, there are a number of fascinating questions in these areas associated with the allocation of time, energy and other resources to yield the most effective identification.

BIBLIOGRAPHY AND COMMENTS

Section 1. Additional discussions of mathematical models and model making are found in the following books:

i. R. Bellman, Adaptive Control Processes: A Guided Tour, (Princeton Univ. Press, Princeton, New Jersey, 1961).

ii. R. Bellman, and C.P. Smith, Simulation in Human Systems - Decision-Making in Psychotherapy, (John Wiley & Sons, Inc., New York, 1973).

iii. R. Bellman, and P. Brock, "On the Concepts of a Problem and Problem-Solving", American Mathematical Monthly, 67, No. 2, pp. 119-134 (1960).

iv. R. Bellman, "The Roles of a Mathematician in Applied Mathematics", Proc. Minneapolis Symp. (1966), pp. 195-204.

Chapter II

QUASILINEARIZATION

1. Introduction

In this chapter we shall develop the mathematical foundations which will be used throughout the book in the study of the methods of systems identifications. We intend to concentrate all the detailed mathematical analysis in this section of the book to avoid repetitions later on.

Central to the development of methods in systems identification as we shall use it, is the quasilinearization technique which shall be introduced in this chapter. By quasilinearization we mean an analytical technique first introduced by Bellman. If we consider the differential equation

$$(1.1) \qquad \frac{dx}{dt} = f(x(t), t) \quad ,$$

the quasilinear technique constructs a sequence of functions $x^{(n)}(t)$ which converges to the solution of Eq. (1.1) under very general conditions. The construction of the sequence of functions is, in many respects, a generalization of the Newton-Raphson method for the extraction of a root of the equation $f(t) = 0$.

In this chapter we shall exploit some of the results of the quasilinear method. The two properties which we shall consider in detail are those of quadratic convergence and monotone sequences. We shall show that the technique not only provides a linear approximate solution to (1.1) but also provides a sequence of functions which in general converges quadratically to the solution of the original nonlinear Eq. (1.1). If a rough initial approximation can be found, then the quasilinear technique provides a sequence of approximate solutions leading to the solution of the original nonlinear Eq. (1.1).

These two properties, linearity and quadratic convergence, lead to a rich tapestry of analytical techniques which form the basis of our methods of systems identification described throughout the book.

2. Newton-Raphson Method

Since the quasilinear technique is motivated by the Newton-Raphson method, we shall review the method briefly. Consider the problem of finding a sequence of approximations to the root of the scalar equation

$$f(x) = 0 \quad . \tag{2.1}$$

We shall assume $f(x)$ is monotone decreasing for all x and strictly convex, i.e., $\frac{d^2f(x)}{dx^2} > 0$, as shown in Fig. 2.1. Hence the root, x_0 is simple, $f(x_0) = 0$.

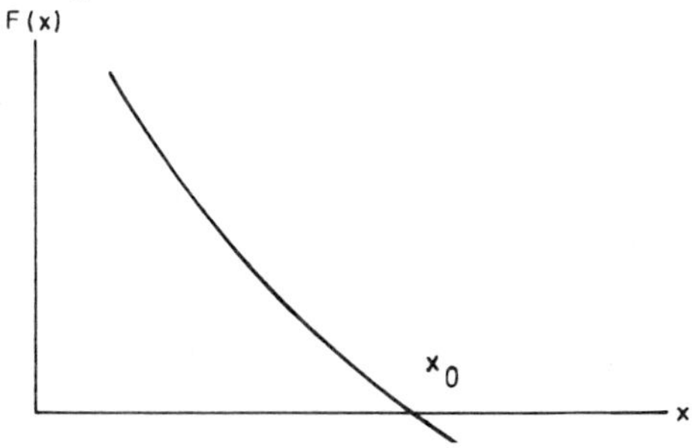

Fig. 2.1

Suppose we have an initial approximation to the root, x_1, such that $x_1 < x_0$ $(f(x_1) > 0)$ and let us expand $f(x)$ about x_1 in a Taylor series (see Fig. 2.2).

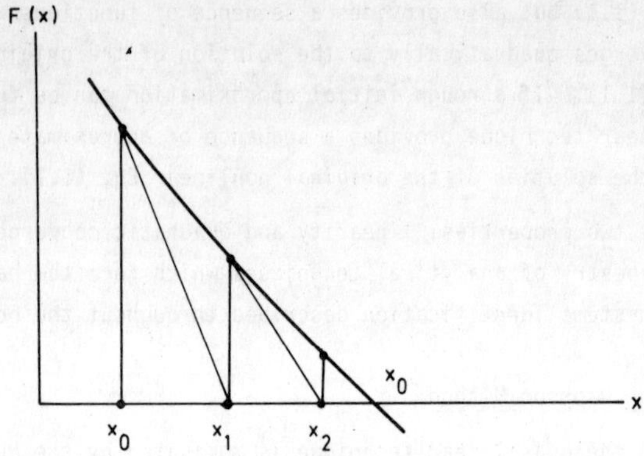

Fig. 2.2

$$f(x) = f(x_1) + (x - x_1)\, f'(x_1) + \ldots \quad . \tag{2.2}$$

A second approximation to x_0 can be obtained by setting $f(x) = 0$ and dropping the higher order terms. We then solve the linear equation for x,

$$f(x_1) + (x_2 - x_1) f'(x_1) = 0 \quad . \tag{2.3}$$

This yields the second approximation,

$$x_2 = x_1 - f(x_1)/f'(x_1) \quad . \tag{2.4}$$

If the process is repeated, the general recurrence relation is

$$x_{n+1} = x_n - f(x_n)/f'(x_n) \quad . \tag{2.5}$$

We immediately note that Eq. (2.5) does not hold if $f'(x_n) = 0$ and that because of monotonicity,

(2.6) $\quad f(x_n) > 0 \;, \qquad f'(x_n) < 0 \quad ,$

the sequence of values generated by (2.5) is monotonically convergent,

(2.7) $\quad x_1 < x_2 < x_3 < \ldots < x_n < x_0 \quad .$

We note here that if the two inequalities (2.6) are not satisfied the monotone convergence property may not be true.

A second important property which we will consider is what we shall call quadratic convergence. By quadratic convergence we mean that for a sequence of approximations, the following inequality is true,

(2.8) $\quad |x_{n+1} - x_0| \le k|x_n - x_0|^2 \quad ,$

where k is independent of x_n. It can be shown that this property holds for the Newton-Raphson technique.

3. Quasilinearization

The quasilinear technique, which will be used throughout this book, is a generalization of the Newton-Raphson method developed in the last section. As such we expect and will indeed show that the properties of monotonicity and quadratic convergence will apply to the quasilinear technique.

We shall begin by considering the nonlinear differential equation,

(3.1) $\quad \frac{dx}{dt} = f(x) \;, \qquad x(0) = c, \qquad t > 0 \quad .$

We shall assume f to be continuous in both x and t, to have continuous bounded second partial derivatives with respect to x for all x and t in the region of interest, and to be strictly convex in x.

The function f can now be expanded around the function $x^{(0)}(t)$ in a Taylor series,

(3.2) $$f(x) = f(x^{(0)}) + \frac{\partial f}{\partial x}(x^{(0)})(x - x^{(0)}) + \ldots$$

where the higher terms have been omitted. The expression $\partial f/\partial x$ represents the partial differentiation of the functional f with respect to the function x and "evaluated" at the function $x^{(0)}(t)$. Combining Eqs. (3.1) and (3.2), we obtain the linear differential equation,

$$\frac{dx}{dt} = f(x^{(0)}) + \frac{\partial f}{\partial x}(x^{(0)})(x - x^{(0)}) ,$$

(3.3) $$x(0) = c .$$

Continuing in this way, we are led to a sequence of continuous functions determined by the recurrence relation,

$$\frac{dx^{(n+1)}}{dt} = f(x^{(n)}) + \frac{\partial f(x^{(n)})}{\partial x}(x^{(n+1)} - x^{(n)}) ,$$

(3.4) $$x^{(n+1)}(0) = c .$$

It is important to note that the recurrence relation is in the form of a linear differential equation requiring an initial approximation $x^{(0)}(t)$ to begin the iterative process. This property will have a profound effect on later developments.

The basic ideas of the quasilinear technique are encompassed in Eq. (3.4). We must emphasize that the success of the technique depends on the initial approximation $x^{(0)}(t)$ and we feel this is so important that we shall devote an entire chapter to determining the initial approximation, $x^{(0)}(t)$.

4. Existence

After having shown that the solution to a nonlinear differential equation can be approximated by a sequence of continuous functions

$x^{(n)}(t)$, we will now be concerned with demonstrating that the sequence does, in fact, exist and furthermore there is a common interval $(0,T)$ in which the functions are uniformly bounded.

We shall begin by defining a norm over the space of all functions $x^{(n)}(t)$ as,

$$\|x^{(n)}(t) - c\| = \max_{0 \le t \le T} |x^{(n)}(t) - c| \quad . \tag{4.1}$$

Under the assumption concerning the boundedness of $f(x)$ and its first two derivatives over a finite interval, we can write

$$\max_{|x-c| < 1} (|f(x)|, |\partial f(x)/\partial x|, \tfrac{1}{2}|\partial^2 f/\partial x^2|) \le m < \infty \quad . \tag{4.2}$$

Let $x^{(0)}(t)$ be selected so that,

$$\|x^{(0)}(t) - c\| < 1 , \qquad 0 \le t \le T \quad . \tag{4.3}$$

Integrating Eq. (3.3) in the last section, we have,

$$x(t) - c = \int_0^t \left[f(x^{(0)}(s)) + (x(s) - x^{(0)}(s)) \frac{\partial f(x^{(0)})}{\partial x} \right] ds \quad . \tag{4.4}$$

From Eqs. (4.2) and (4.3), it follows that,

$$\|x(t) - c\| < t \left[m + m(\|x(t) - c\| + 1) \right] \quad . \tag{4.5}$$

so that for $0 \le t \le T$,

$$\|x(t) - c\| \le T \left[m + m(\|x(t) - c\|) \right] \quad , \tag{4.6}$$

or,

$$\|x(t) - c\| \le \frac{2Tm}{(1 - Tm)} \tag{4.7}$$

but $\| x(t) - c \| \le 1$, only if

$$0 \le t \le T < \frac{1}{3m} \quad . \tag{4.8}$$

We now proceed inductively, under the assumption,

$$\| x(t) - c \| < 1, \qquad 0 \le t \le T \quad . \tag{4.9}$$

We find, as before,

$$x^{(n+1)}(t) - c = \int_0^t f\left(x^{(n)}(s)\right) + \left(x^{(n+1)}(s) - x^{(n)}(s)\right) \frac{\partial f(x^{(n)})}{\partial x} ds \quad , \tag{4.10}$$

hence we have,

$$\| x^{(n+1)}(t) - c \| < \frac{2Tm}{(1 - Tm)}$$

which will not exceed unity provided (4.8) holds. Thus we see that under the boundedness condition (4.2), we know a region exists in which the sequence of functions $x^{(n)}(t)$ generated by the quasilinear technique is uniformly bounded.

5. Convergence

A very important property of the sequence $x^{(n)}(t)$ is that of convergence. We say a sequence of functions $x^{(n)}(t)$ is quadratically convergent if it converges to a limit function $x(t)$, and

$$\| x^{(n+1)}(t) - x(t) \| \le k \, \| x^{(n)}(t) - x(t) \|^2 \tag{5.1}$$

where k is independent of n.

To show the conditions under which the sequence $x^{(n)}(t)$ is quadratically convergent, we consider the quasilinear system,

$$\frac{dx}{dt} = f(x) , \qquad x(0) = c, \qquad 0 \le t \le T , \tag{5.2}$$

$$\frac{dx^{(n+1)}}{dt} = f(x^{(n)}) + \frac{\partial f(x^{(n)})}{\partial x} (x^{(n+1)} - x^{(n)}) , \tag{5.3}$$

$$\frac{dx^{(n)}}{dt} = f(x^{(n-1)}) + \frac{\partial f(x^{(n-1)})}{\partial x} (x^{(n)} - x^{(n-1)}) . \tag{5.4}$$

Using Eqs. (5.2) and (5.4), we may write,

$$\frac{d}{dx} (x - x^{(n)}) = f(x) - f(x^{(n)}) - \frac{\partial f(x^{(n-1)})}{\partial x} (x^{(n)} - x^{(n-1)}) ,$$

$$0 \le t \le T , \qquad n = 0,1,2,\dots . \tag{5.5}$$

Rewriting (5.5), we have

$$\frac{d}{dx} (x - x^{(n)}) = f(x) - f(x^{(n)}) + \frac{\partial f(x^{(n-1)})}{\partial x} (x^{(n-1)} - x)$$

$$- \frac{\partial f(x^{(n-1)})}{\partial x} (x^{(n)} - x) . \tag{5.6}$$

Integrating (5.6) from 0 to t and since $x(0) - x^{(n)}(0) = 0$, we have

$$x - x^{(n)} = \int_0^t \left[(f(x) - f(x^{(n)}) - \frac{\partial f(x^{(n-1)})}{\partial x} (x - x^{(n-1)}) \right.$$

$$\left. - \frac{\partial f(x^{(n-1)})}{\partial x} (x^{(n)} - x) \right] dt . \tag{5.7}$$

Applying the mean value theorem,

$$x - x^{(n)} = \int_0^t \left(\frac{1}{2} \frac{\partial^2 f(x^*)}{\partial x^2} (x - x^{(n-1)})^2 + \frac{\partial f(x^*)}{\partial x} (x^{(n)} - x) \right) dt . \tag{5.8}$$

where $x^*(t)$ lies between $x^{(n-1)}(t)$ and $x^{(n)}(t)$ in the interval, $0 \le t \le T$.

Using the bounds already established,

$$(5.9) \qquad \| x - x^{(n)} \| \le T(m \| x - x^{(n)} \|^2 + m \| x - x^{(n)} \|) ,$$

giving the relation

$$(5.10) \qquad \| x - x^{(n)} \| \le \frac{Tm}{(1 - Tm)} \| x - x^{(n-1)} \|^2 .$$

Therefore, the sequence converges quadratically and since T is chosen such that $Tm \le 1/3$, the sequence converges under the assumption $x(t)$ is the solution of (5.3).

6. Discussion

Since the quasilinear technique is a fundamental concept which we shall use throughout the book, we shall pause to discuss what we have established.

As we have shown, the quasilinear technique defines a sequence of functions which converge to the solution of a known differential equation. The essential property of this sequence of functions $x^{(n)}(t)$ is that each is defined by a linear differential equation with time varying coefficients. Later in this chapter we shall exploit this property.

The other important properties of the sequence $x^{(n)}(t)$ are those of quadratic convergence and monotonicity. The first is an immediate consequence of the generalized Newton-Raphson approximation. Like the classical Newton-Raphson method, the quasilinearization technique requires a first approximate solution which is close to the solution. This fact will be exploited in greater depth in a later chapter.

The property of monotonicity holds only for special classes of operators which have the required properties of positivity and convexity.

Yet theoretically we can always make the differential operator possess these properties by making the interval $(0,T)$ sufficiently small.

It seems from our study that the required properties of positivity and convexity would be a severe limitation on the usefulness of the quasilinear technique. Yet in the examples shown throughout this book this is not the case and although the convexity and positivity properties play an important role in the derivation of the quasilinear technique, they do not seem to arise in practical problems.

The most exploited property of the quasilinear technique is its ability to solve nonlinear problems as a sequence of linear ones. In the ensuing chapters we shall be constantly referring to the quasilinear technique in considering nonlinear models and we must continually note that the properties discussed here will always be hidden from view but are everpresent.

7. Quasilinear Equations

The quasilinear equations developed in Sec. 3, namely,

$$\frac{dx^{(n+1)}}{dt} = f(x^{(n)}) + \frac{\partial f(x^{(n)})}{\partial x}(x^{(n+1)} - x^{(n)}) \tag{7.1}$$

possesses the property of linearity, that is, if $g_1(t)$ and $g_2(t)$ are solutions of (7.1), then so is,

$$g(t) = g_1(t) + g_2(t) \quad . \tag{7.2}$$

This property will play a fundamental role in our development of the methods in the identification of systems and we shall consider its impact in a later part of this chapter.

8. Systems of Differential Equations

Let us consider a system of N first order nonlinear differential equations of the form,

$$\text{(8.1)} \qquad \frac{dx_i}{dt} = f_i(x_1, x_2, \ldots, x_N, t) \;, \qquad i = 1, 2, \ldots, N \;,$$

with the initial conditions,

$$\text{(8.2)} \qquad x_i(0) = c_i \;.$$

If this system is written in a general vector form,

$$\text{(8.3)} \qquad \frac{dX}{dt} = F(X,t) \;,$$

where X and F are N-dimensional vectors, then the quasilinear technique yields a general recursion relation in the same way as was done for a single differential equation,

$$\text{(8.4)} \qquad \frac{dX^{(n+1)}}{dt} = F(X^{(n)}) + J(X^{(n)})(X^{(n+1)} - X^{(n)}) \;,$$

where $J(X^{(n)})$ is the Jacobi matrix,

$$\text{(8.5)} \qquad J(X^{(n)}) = \begin{bmatrix} \frac{\partial f_1(X^{(n)})}{\partial x_1} & \frac{\partial f_1(X^{(n)})}{\partial x_2} & \cdots & \frac{\partial f_1(X^{(n)})}{\partial x_N} \\ \frac{\partial f_2(X^{(n)})}{\partial x_1} & \frac{\partial f_2(X^{(n)})}{\partial x_2} & \cdots & \frac{\partial f_2(X^{(n)})}{\partial x_N} \\ \cdot & \cdot & & \cdot \\ \cdot & \cdot & & \cdot \\ \cdot & \cdot & & \cdot \\ \frac{\partial f_N(X^{(n)})}{\partial x_1} & \frac{\partial f_N(X^{(n)})}{\partial x_2} & \cdots & \frac{\partial f_N(X^{(n)})}{\partial x_N} \end{bmatrix} .$$

The elements of the Jacobi matrix $J(X^{(n)})$ are functions of the particular variables $x_i(t)$ and we use the n-th iterative solution in the evaluation of $J(X^{(n)})$.

We shall now turn our attention to the numerical solution of the system defined by Eq. (8.4). We shall assume for the moment that an initial approximate solution $X^{(0)}(t)$ is known and $0 < t < T$ is the time interval over which the solutions $X^{(n)}(t)$ are uniformly convergent.

If the initial conditions of the system (8.4) are known, then by substituting the initial approximation $X^{(0)}(t)$ into Eq. (8.4), the resulting initial value problem can be integrated numerically without difficulty over the range $0 < t < T$. By repeating the process for every new iteration, a sequence of solutions can be generated which would converge to the solution of (8.1).

The quasilinearization technique becomes a powerful method in the identification of systems when the initial conditions are not known. In this situation the initial conditions must be identified. A successful method is to use the linear property of the system of Eqs. (8.4), that is, if $Y_1(t)$ and $Y_2(t)$ are two vector solutions of (8.4), then

$$(8.6) \qquad Y(t) = a\, Y_1(t) + b\, Y_2(t)$$

is also a solution.

Using this property we observe that at each iteration the complete solution to the vector quasilinear Eq. (8.4) can be written as the sum of a particular plus n homogeneous solution of (8.4),

$$(8.7) \qquad X^{(n+1)}(t) = P^{(n+1)}(t) + \sum_{i=1}^{n} \alpha_i H_i^{(n+1)}(t) \quad ,$$

where $P^{(n+1)}(t)$ is the particular vector solution to Eq. (8.4) with the initial conditions $P^{(n+1)}(0) = 0$ and $H_i^{(n+1)}(t)$ is the i-th homogeneous vector solution to (8.4) with the initial conditions,

$$H_i^{(n+1)}(0) = \delta_{ij} , \qquad j = 1,2,\ldots,N , \tag{8.8}$$

where j defines the j-th component of the vector $H_i^{(n+1)}$.

By defining initial conditions for both the particular and N homogeneous vector equations, these solutions can be obtained numerically simply by integrating the equations over the time interval $(0,T)$. A close inspection of Eq. (8.8) together with the initial conditions, shows that the coefficients α_i are precisely the unknown system initial conditions.

This is a very important observation for now the identification problem is reduced to the determination of N coefficients α_i, $i = 1,2,\ldots,N$. This reduction is possible because the particular and homogeneous solutions can be computed numerically independent of the initial conditions of the system.

9. System Parameters and Initial Conditions

Let us now consider a system generated by the equation,

$$\frac{dx(t)}{dt} = f(x(t), a) , \qquad x(0) = c , \tag{9.1}$$

where a parameter, a, has been added to the system.
We wish to determine both the parameter a and the initial condition c, but before we consider the identification problem, we must deal with the parameter a. It is inconvenient to consider the system parameter a and the initial condition c as two different types of scalars. Let us consider a to be a state variable where a is subject to the differential equation,

$$\frac{da(t)}{dt} = 0 . \tag{9.2}$$

Then the identification problem requires an examination of the augmented system,

$$\frac{dx(t)}{dt} = f[x(t), a] , \qquad x(0) = c$$

$$(9.3) \qquad \frac{da(t)}{dt} = 0 \qquad , \qquad a(0) = a_0 \quad .$$

The objective of the identification problem then is to determine the initial state variables $a(0)$ and $x(0)$ in such a way as to minimize some penalty function which is discussed in the next section.

10. The Penalty Function

The identification problem, as we have defined it, is to determine the initial state vector $a(0)$ and $x(0)$ in such a way that we minimize some penalty function. Although we wish to discuss this topic in detail in a later chapter, we shall introduce here a criterion we have found most useful in practical problems.

Let us assume that the system we are studying is observed over a time interval $(0,T)$ and such observations are denoted by $r(t)$.

We take as a penalty function,

$$(10.1) \qquad J = \int_0^T \left[p^{(n+1)}(t) + \sum_{i=1}^{N} a_i H_i^{(n+1)} - r(t)\right]^2 dt$$

which represents the least squares penalty function. If Eq. (10.1) is interpreted as a vector dot product, then it can be written in the form,

$$(10.2) \qquad J = \frac{1}{2} a^T A a + B^T a + c$$

where A is a positive definite matrix, B is a vector and c a scalar constant. The problem now is to determine the vector a^* which will minimize J. High among the numerical techniques is the conjugate gradient method which minimizes J in N iterations or less.

Recalling the quasilinear technique, the set of scalars a_i(vector a) which minimizes J is an approximate solution to the identification problem and this knowledge allows us to continue the iterative process until acceptable convergence is achieved.

11. The Initial Approximation

As we alluded to in Sec. 5, the initial approximate solution is very important if convergence is to be assured. While we shall be discussing this problem in detail in a later chapter, now we only wish to indicate several ways of obtaining this initial approximation.

A neat way to obtain the first approximation is to select, by an educated guess, a set of initial conditions c^*, a^* for the system,

$$\frac{dx(t)}{dt} = f(x(t), a), \quad x(0) = c^* ,$$

$$\frac{da(t)}{dt} = 0 \quad , \quad a(0) = a^* ,$$

$$0 < t < T . \tag{11.1}$$

We simply integrate the nonlinear system (11.1) numerically from $t=0$ to $t=T$. The resulting function $x^*(t)$, a^* now becomes the initial approximation for the quasilinear technique.

If the observed data $r(t)$ is known close to the initial time $t=0$, then an approximate set of initial conditions can be found by extrapolating the data to $t=0$.

Finally it would seem sensible to use the observed data itself as the first approximation, thus avoiding the necessity of integrating the nonlinear equation (11.1) at all. Of course this technique requires that $r(t)$ be known as a continuous function of t. If discrete observations are made, then interpolation schemes, such as the cubic spline, can be used to obtain a continuous approximation to the data.

12. Discussion

We have shown in this chapter how the identification problem could be defined precisely and solved easily within the structure of the quasilinear technique. It became apparent from an examination of the method that the quasilinear method led to the construction of a series of functions which, if they converged at all, converged quadratically, and there could always be found a neighborhood about the origin where this convergence is monotone.

In a large majority of the examples which we shall be studying throughout the book, we shall use the quasilinear technique. It is our belief that a well stated foundation for the quasilinear technique, once stated, can serve as the basis for its use throughout the book.

BIBLIOGRAPHY AND COMMENTS

1. Section 1. A more detailed discussion of quasilinearization and other examples is found in the book by R. Bellman and R. Kalaba, Quasilinearization and Nonlinear Boundary-Value Problems, (American Elsevier Pub. Co. Inc., New York, 1965).

2. A more theoretical discussion is found in the book by R. Bellman, Methods of Nonlinear Analysis, Vol II, (Academic Press, Inc., New York, 1973).

3. Other applications are found in the following papers:

   i. R. Bellman, "Quasilinearization and Upper and Lower Bounds for Variational Problems", Quarterly of Applied Mathematics, 19, 349-350 (1962).

   ii. R. Bellman, "Dynamic Programming, Invariant Imbedding, and Quasilinearization: Comparison and Interconnections", Computational Methods in Optimization Problems, (Academic Press, Inc., New York, 1964), pp. 135-145.

   iii. R. Bellman, R. Kalaba and R. Sridhar, "Adaptive Control via Quasilinearization and Differential Approximation", Pakistan Engineer, 5, No. 2, 94-100 (1965).

   iv. R. Bellman, H. Kagiwada and R. Kalaba, "Quasilinearization, Boundary-Value Problems and Linear Programming", IEEE Automatic Control, AC-10, No. 2, 199 (1965).

   v. R. Bellman, H. Kagiwada and R. Kalaba, "Wengert's Numerical Method for Partial Derivatives, Orbit Determination, and Quasilinearization", Comm. Assoc. for Computing Machinery, 8, No. 4, 231-232 (1965).

   vi. R. Bellman, H. Kagiwada and R. Kalaba, "Quasilinearization, System Identification, and Prediction", International Journal of Engineering Science, 3, 327-334 (1965).

vii. R. Bellman, H. Kagiwada and R. Kalaba, "Quasilinearization and the Calculus of Eigenvalues", *Comm. Assoc. for Computing Machinery*, *9*, 522-524 (1966).

viii. R. Bellman, H. Kagiwada and R. Kalaba, "Quasilinearization and the Estimation of Time Lags", *Math. Biosci.* *1*, No. 1, 39-44 (1967).

ix. R. Bellman, J. Jacquez, R. Kalaba and J. Schwimmer, "Quasilinearization and the Estimation of Chemical Rate Constants from Raw Kinetic Data", *Math. Biosci.*, *1*, No. 1, 71-76 (1967).

x. R. Bellman, H. Kagiwada and R. Kalaba, "Quasilinearization and the Estimation of Differential Operators from Eigenvalues", *Comm. Assoc. for Computing Machinery*, *11*, No. 4, 255-256 (1968).

xi. R. Bellman, "Quasilinearization and Stability of Nonlinear Stochastic Systems", *J. of Math. Anal. Appl.*, *40*, No. 1, 159-162 (1972).

xii. R. Bellman and R. Vasudevan, "Quasilinearization and the Matrix Riccati Equation", *J. of Math. Anal. Appl.*, *64* (1978).

4. Other methods of systems identification are given in the following papers:

i. C.B. Neal and G.A. Bekey, "Estimation of the Parameters of Sampled-Data Systems by Means of Stochastic Approximation", *Int. J. Systems Sci.*, *1*, No. 1, 63-74 (1970).

ii. G.O. Burnham, J. Seo, and G.A. Bekey, "Identification of Human Driver Models in Car Following", *IEEE Transactions on Automatic Control*, *AC-19*, No. 6, 911-915 (Dec. 1974).

iii. G.A. Bekey and S.M. Yamashiro, "Parameter Estimation in Mathematical Models of Biological Systems", *Advances in Biomedical Engineering*, *6*, 1-43 (1976).

iv. J.C. Maloney Jun and G.A. Bekey, "Online Estimation of Cardiac Output from Respiratory Data using a Gradient Method", *Medical and Biological Engineering*, pp. 379-386 (July 1976).

v. G.A. Bekey, "Parameter Estimation in Biological Systems: A Survey", *In Identification and System Parameter Estimation*, Proc. 3rd IFAC Symp. edited by P. Eykhoff, (North-Holland/ American Elsevier, Amsterdam, 1973).

vi. T.M. Grove and G.A. Bekey, "Sensitivity Problems in the Identification of Biological System Parameters", *Proc. Symp. on Appl. of Computer Methods in Engineering*, Los Angeles, August 1977.

vii. G.A. Bekey and J.E.W. Beneken, "Identification of Biological Systems: A Survey", Automatica, 14, 41-47 (1978).

5. Some discussions of observation and experimentation are found in the book by R. Bellman, Mathematical Methods in Medicine, (World Scientific Pub. Co., Singapore, 1983).

Chapter III

ORBIT DETERMINATION

1. Introduction

The basic ideas underlying the solution of problems via quasilinearization are discussed in Chapter II. An example of quasilinearization is described in this chapter as an application to the computational solution of some general N-point boundary-value problems and to present an application to orbit determination.

2. An N-Point Boundary-Value Problem

Consider an N-dimensional vector $x(t) = (x_1(t), x_2(t),\ldots,x_N(t))$ which is a solution of the vector system of equations

$$\frac{dx}{dt} = f(x,t) \ , \qquad 0 \le t \le b \ , \tag{2.1}$$

and in addition satisfies the N conditions

$$\sum_{j=1}^{N} a_j(t_i)x_j(t_i) = r_i \ , \qquad i = 1,2,\ldots,N \ , \tag{2.2}$$

where the coefficients $a_j(t_i)$ and r_i are known and

(2.3) $$0 \le t_1 \le t_2 \le \dots \le t_N \le b \ .$$

Let us assume that Eqs. (2.1) and (2.2) possess a unique solution on the interval $[0,b]$. Our objective is to provide an efficient computational algorithm for determining the function $x(t)$ on the interval $[0,b]$.

3. Solution via Quasilinearization

Let $x^{(0)}(t)$ be an initial approximation on the given interval. The (k+1)st approximation, $x^{(k+1)}(t)$, is to be obtained from the k-th approximation $x^{(k)}(t)$, through use of the relations

(3.1) $$\frac{dx_i}{dt} = f_i(x^{(k)},t) + \sum_{j=1}^{N} (x_j^{(k+1)} - x_j^{(k)}) \frac{\partial f_i(x^{(k)},t)}{\partial x_j} \ ,$$

$$i = 1,2,\dots,N \quad \text{and} \quad k = 0,1,2,\dots$$

with the boundary conditions

(3.2) $$\sum_{j=1}^{N} a_j(t_i) x_j^{(k+1)}(t_i) = b_i \ , \qquad i = 1,2,\dots,N \ ,$$

for $k = 0,1,2,\dots$. The linear system for $x^{(k+1)}$, Eq. (3.1), subject to the linear conditions of Eq. (3.2), is to be solved in the following manner. We determine a particular solution of Eq. (3.1) and N independent homogeneous solutions via numerical integration. This is readily accomplished since these are all initial-value problems. Then $x_i^{(k+1)}(t)$ is expressed as the sum of a particular and n homogeneous solution of (3.1).

(3.3) $$x_i^{(k+1)}(t) = p_i^{(k+1)}(t) + \sum_{j=1}^{N} a_j h_{j,i}^{(k+1)}(t) \ ,$$

using an obvious notation. The N unknown constants, a_j, are then determined by solving the linear algebraic equations obtained by substituting the expression in (3.3) by Eq. (3.2).

In this way the solution is affected by reducing the original problem to a sequence of linear initial-value problems, a process well suited for digital computer calculations. Some aspects of convergence (which is quadratic in nature) for the case of two-point boundary-value problems have been discussed in Chapter II.

4. Orbit Determination

Consider the problem of determining the unperturbed orbit of a heavenly body H in the (x,y) plane, the plane of the ecliptic, on the basis of four angular observations made from the earth (see Fig. 3.1).

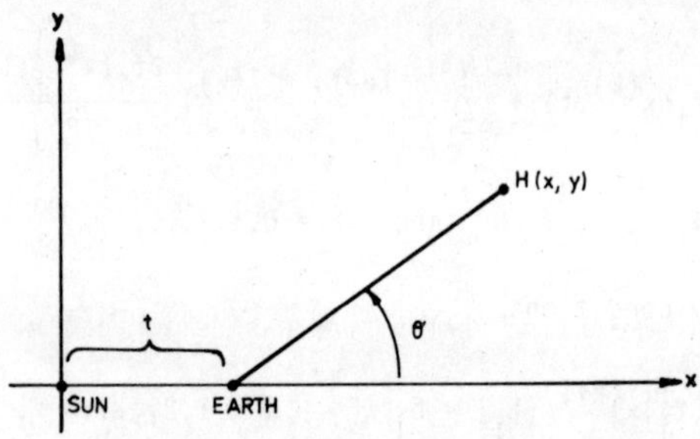

Fig. 3.1 The Physical Situation

Assume that the units are normalized so that the equations of motion of H are

$$\frac{dx}{dt} = u, \qquad \frac{du}{dt} = -\frac{x}{(x^2 + y^2)^{1.5}} ,$$

$$\frac{dy}{dt} = v, \qquad \frac{dv}{dt} = -\frac{y}{(x^2 + y^2)^{1.5}} . \tag{4.1}$$

At times t_i, $i = 1,2,3,4$, we are given that

$$\theta(t_i) = \theta_i , \tag{4.2}$$

which yields the conditions

$$y(t_i) = (x(t_i) - 1) \tan \theta_i \ , \qquad i = 1,2,3,4 \ . \tag{4.3}$$

We wish to determine a complete set of elements of the orbit in the form of values for x, u, y, and v for some convenient time t. For simplicity we assume that the earth is stationary and that the sole force on the heavenly body is the gravitational attraction of the sun.

5. Numerical Results

Consider the case for which the observational data are (in radians):

$$\theta(0.5) = 0.251297 \ , \qquad \theta(1.0) = 0.510240 \ ,$$

$$\theta(1.5) = 0.783690 \ , \qquad \theta(2.0) = 1.076540 \ . \tag{5.1}$$

These data were generated by assuming that at time zero we have $x(0) = 2.0$, $y(0) = 0$, $u(0) = 0$, $v(0) = 0.5$. The situation, with a stationary earth, is as shown in Fig. 3.2. As an initial approximation to the trajectory, we assume that the observed body is at the point $(1.0, 0)$ at time zero and that it moves in a circular orbit with a period equal to 3.664, which is a very poor initial approximation.

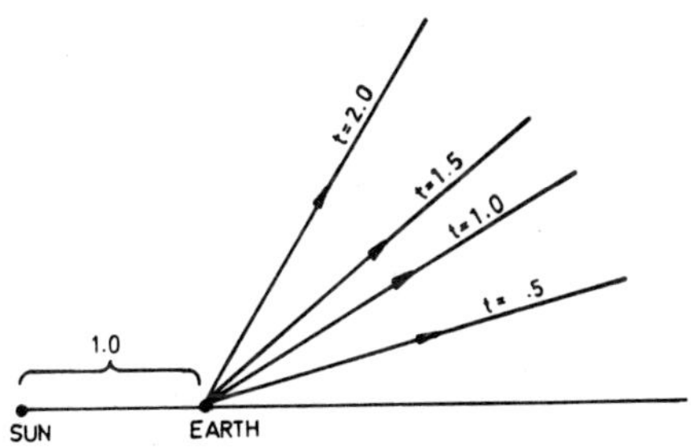

Fig. 3.2 Approximate Directions to the Heavenly Body for Four Angular Observations from a Stationary Earth

Then using a Runge-Kutta integration method and the procedure sketched above, we found the following:

TABLE I. The Approximations to the Displacements and Velocities at Time 2.5

	Initial approx.	First approx.	Second approx.	Third approx.	Fourth approx.	Precise value to 6 figures
x(2.5)	-0.455175	1.554360	1.169660	1.190170	1.193560	1.193610
u(2.5)	-0.873320	-1.638830	-0.495101	-0.656686	-0.664207	-0.664263
y(2.5)	0.975790	2.982690	0.865184	1.043930	1.060540	1.060700
v(2.5)	-0.408690	1.807620	0.201206	0.237695	0.247433	0.247499

It produced, of course, not merely the results of Table I but an ephemeris for $0 \le t \le 2.5$ at intervals of 0.01.

Perhaps even more surprising is the fact that if, as an initial approximation to the orbit, we consider the heavenly body coincides with the position of the earth over all time we still obtain the same rapid convergence to the correct orbit.

6. Discussion

The method is readily extended to cases where more than four observations are available through use of the method of least squares. Furthermore, perturbations due to the motion of the earth and the presence of other planets are readily incorporated, since they merely result in modifications of Eq. (4.1). Of course, closely spaced observations present difficulties in the numerical determination of the constants, a_j of Eq. (3.3).

BIBLIOGRAPHY AND COMMENTS

Section 1. We are following the paper by R. Bellman, H. Kagiwada, and R. Kalaba, "Orbit Determination as a Multi-Point Boundary Value Problem and Quasilinearization", Proc. Nat. Aca. Sci., 48, No. 8, 1327-1329 (August 1962).

Chapter IV

ESTIMATION OF CHEMICAL RATE CONSTANTS

1. Introduction

In this chapter we will indicate the application of the general method of quasilinearization to the estimation of chemical rate constants using raw kinetic data. Bodenstein and Lindner have presented data on pressure observed during the reversible homogeneous gas phase reaction,

$$2NO + O_2 \rightleftharpoons 2NO_2 \quad . \tag{1.1}$$

The experiments were carried out at constant volume and temperature. The forward reaction is third order and the backward reaction is second order. The descriptive differential equation, after various normalizations, has the form,

$$\frac{dx}{dt} = k(a-x)(b-x)^2 - Kx^2 \quad , \tag{1.2}$$

where $a = 126.2$ and $b = 91.9$, and k and K are the rate constants to be estimated on the basis of the observations given in Table I.

Let us denote the n-th observation at time t_n by w_n, $t_1 = 1$, $w_1 = 1.4$, and so on. Next, we have to determine the rate constants k and K so that the solution of the differential Eq. (1.2), subject to

TABLE I. Experimental Data

t	0	1	2	3	4	5	6	7	9	11	14	19	24	29	30
x	0	1.4	6.3	10.5	14.2	17.6	21.4	23.0	27.0	30.5	34.4	48.8	41.6	43.5	45.3

the initial condition $x(0) = 0$, will be as close as possible to the measured values. We must, of course, introduce an appropriate measure of closeness.

2. Quasilinearization

We can first transform the problem of estimating constants within the equation into one which we estimate the initial conditions for a system of differential equations. We do this by considering k and K to be functions of time that satisfy the differential equations

$$\frac{dk}{dt} = 0 \quad , \tag{2.1}$$

$$\frac{dK}{dt} = 0 \quad . \tag{2.2}$$

The aim is to consider all quantities as functions of time, and all unknowns as initial conditions.

As a measure of closeness, we now minimize the quantity

$$Q = \sum_{i=1}^{M} (x(t_i) - w_i)^2 \quad , \tag{2.3}$$

where

$$\frac{dx}{dt} = k(126.2 - x)(91.9 - x)^2 - Kx^2 \; , \qquad x(0) = 0 \quad , \tag{2.4}$$

$$\frac{dk}{dt} = 0 \quad , \tag{2.5}$$

$$\frac{dK}{dt} = 0 \quad . \tag{2.6}$$

The minimization is over the initial values of k and K, $k(0)$ and $K(0)$.

We will sketch the quasilinearization technique in quite general terms. Let the N-dimensional vector $X(t)$ be a solution of the differential equation

$$\frac{dx}{dt} = f(x) ,$$

$$x_i(0) = \alpha_i , \qquad i = 1,2,..., \tag{2.7}$$

with initial conditions on each component.
The first S components of $x(0)$ are specified, and the last $N-S$ are free to be chosen so as to minimize the sum

$$Q = \sum_{i=1}^{M} [(\beta, x(t_i)) - w_i]^2 , \qquad M \geq N - S . \tag{2.8}$$

As usual, (β,x) denotes an inner product, and β is a constant weighting vector. As was remarked earlier, our procedure is an iterative one. First we select an initial approximation to the missing initial conditions and integrate Eq. (2.7) numerically on the interval $0 \leq t \leq t_m$. Call this vector function $X^0(t)$. Next we consider a linearized version of Eq. (2.7),

$$\frac{dx^1}{dt} = f(X^0) + J(X^0)(X^1 - X^0) , \tag{2.9}$$

where the Jacobian matrix J is given by

$$J(X^0) = \left[\frac{\partial f_i}{\partial x_j} \right] . \tag{2.10}$$

This matrix is known computationally on the interval $0 \leq t \leq t_m$. We produce numerically a particular solution $p(t)$ of Eq. (2.9) on the interval $0 \leq t \leq t_m$, using some convenient initial conditions, such as

$$(2.11)\qquad P_i(0) = \begin{cases} \alpha_i\ , & i = 1,2,\ldots,S\ , \\ 0\ , & i = S+1,\ldots,N\ . \end{cases}$$

Then we produce numerically $N-S$ independent vector solutions of the homogeneous equation

$$(2.12)\qquad \frac{dH_j}{dt} = J(X^0)H_j\ , \qquad j = S+1,\ S+2,\ldots,N\ ,$$

where $H_j(t)$ is an N-dimensional vector. For initial conditions we choose

$$(2.13)\qquad H_j(0) = \delta_{ij}\ , \qquad j = S+1,\ S+2,\ldots,N\ .$$

Then the solution of Eq. (2.9) is representable in the form

$$(2.14)\qquad X^1(t) = P(t) + \sum_{j=S+1}^{N} c_j H_j(t)\ ,$$

where $c_{S+1}, c_{S+2},\ldots,c_N$ are constants to be determined. They are calculated as solutions of the linear algebraic equations

$$(2.15)\qquad \frac{\partial Q}{\partial c_j} = 0\ , \qquad j = S+1,\ S+2,\ldots,N\ ,$$

where for $X(t_i)$ we substitute

$$(2.16)\qquad X(t_i) = P(t_i) + \sum_{j=S+1}^{N} c_j H_j(t_i)\ , \qquad i = 1,2,\ldots,M\ .$$

In view of our choices of $p(0)$ and $h_j(0)$, we have as our new approximation the initial vector

$$
(2.17) \qquad X(0) = \begin{bmatrix} \alpha_1 \\ \cdot \\ \cdot \\ \cdot \\ \alpha_S \\ c_{S+1} \\ \cdot \\ \cdot \\ \cdot \\ c_N \end{bmatrix}
$$

The process can then be repeated to obtain an improved estimate.

3. Numerical Results

The quasilinearization technique described in the Introduction was applied to the problem. The results are shown in Table II.

TABLE II. Numerical Results

Approximation	k	K
0	1×10^{-6}	1×10^{-4}
1	0.3413×10^{-6}	0.2554×10^{-2}
2	0.4589×10^{-5}	0.3583×10^{-3}
3	0.4578×10^{-5}	0.2808×10^{-3}
4	0.4577×10^{-5}	0.2797×10^{-3}

By comparison, Bodenstein and Lindner, using a combination of chemical theory and the observations given earlier, estimated that

$$(3.1) \qquad k = 0.53 \times 10^{-5} \quad ,$$

$$(3.2) \qquad K = 0.41 \times 10^{-3} \quad .$$

As a point of interest, the sum of the squares of the deviations for our choices of the parameters is 0.210×10^{-2}, and the sum of the squares of the deviations for Bodenstein's and Lindner's choice is 0.555×10^{-2}.

We do not imply that our values are more accurate than those of Bodenstein and Lindner. We only emphasize that our estimates were obtained in a systematic, computationally feasible manner requiring no further knowledge of the chemical process than that contained in Eq. (1.2) plus the observations, and the initial estimates.

BIBLIOGRAPHY AND COMMENTS

Section 1. We are following the paper by R. Bellman, J. Jacquez, R. Kalaba, and S. Schwimmer, "Quasilinearization and the Estimation of Chemical Rate Constants from Raw Kinetic Data", Math. Biosci., 1, 71-76 (1967).

Chapter V

INVERSE PROBLEMS IN ECOLOGY

1. Introduction

Stimulated by his friend's questions concerning an explanation of the fluctuations of the yearly catch of certain types of fish, Volterra began a systematic study of the interactions of populations. If there are k interacting species under certain simplified assumptions a reasonable description of the process over time can be obtained by the study of a system of ordinary differential eouations

$$\frac{dN_i}{dt} = g_i(N_1, N_2, \ldots, N_k) \quad , \tag{1.1}$$

$N_i(0) = c_i$, $i = 1,2,\ldots,k$. For more precise descriptions, it is necessary to employ differential-difference equations and generally equations with hereditary effects.

Here, we propose to study the problem of determining how accurate an explanation of observed phenomena can be obtained using a system of the type appearing in Eq. (1.1). The technique we shall employ is that of quasilinearization, aided by a digital computer. It is only with the aid of these devices that we can hope to tackle these in a unified and sophisticated fashion.

To illustrate the approach, we shall consider the simple case of two interacting species, that of the prey and the predator. An interesting feature is a series of experiments determining the sensitivity of the parameters in the model to the accuracy of the observations of the population sizes at various times.

2. An Inverse Problem

Let us consider the Lotka-Volterra equations

$$\frac{dN_1}{dt} = (c_3 - c_5 N_2)N_1 , \qquad N_1(0) = c_1 ,$$

$$\frac{dN_2}{dt} = -(c_4 - c_6 N_1)N_2 , \qquad N_2(0) = c_2 , \tag{2.1}$$

where $c_1, c_2, c_3, c_4, c_5, c_6$ are positive constants. The parameters c_3, c_4 correspond to growth constants, while c_5, c_6 represent interaction coefficients; N_1, N_2 denote respectively the prey and predator populations.

Consider the case in which the constants are:

$$\begin{aligned} c_1 &= 1.2 , \\ c_2 &= 1.1 , \\ c_3 &= 1.0 , \\ c_4 &= 1.0 , \\ c_5 &= 1.0 , \\ c_6 &= 1.0 . \end{aligned} \tag{2.2}$$

We produce the population curves as functions of time by integrating Eqs. (2.1), using the constants given by Eq. (2.2), with a step length of $\Delta t = 0.01$. These curves, on the interval $0 \leq t \leq 10$, are shown in Fig. 5.1.

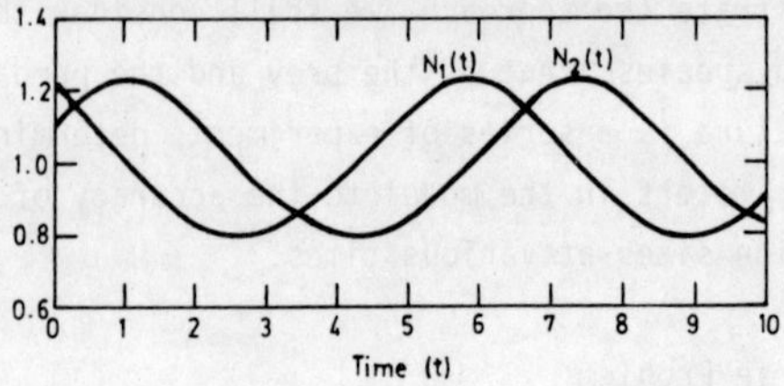

Fig. 5.1 Curves of Population Growth of Two Interacting Species

Now we pose the following inverse problem: Given the observations of N_1, $\{a_i\}$, and the observations of N_2, $\{b_i\}$, at discrete times t_i, $i = 1,2,\ldots,M$, determine the values of the six constants $c_1,c_2,\ldots,c_6$ which would produce solutions of Eqs. (2.1) in best agreement with the observations, in a least squares sense. We wish to minimize the expression

$$(2.3) \qquad \sum_{i=1}^{M} \{[N_1(t_i) - a_i]^2 + [N_2(t_i) - b_i]^2\}$$

by a proper choice of $c_1,c_2,\ldots,c_6$.

This is a nonlinear multipoint boundary-value problem which can be solved via the successive approximation method of quasilinearization. Suppose that there is a current approximation of the solution. Improved approximations are produced by solving a sequence of related linear multipoint boundary-value problems. The sequence of such approximations is quadratically convergent, when convergent at all.

For the ecology problem at hand, we begin with an estimate of each of the six constants $c_1,c_2,\ldots,c_6$, and the two population functions $N_1(t)$ and $N_2(t)$. In order to obtain the next set of approximations, we integrate numerically 14 linearized equations, then invert a 6×6 matrix to get improved values of the six constants, and we use these constants to produce improved approximations of $N_1(t)$ and $N_2(t)$. By repeating these steps several times, we rapidly converge to a solution of the inverse problem.

3. Computational Results

We performed a series of numerical experiments for the determination of the six constants. In some trials, we truncated the correct values $N_1(t_k)$, $N_2(t_k)$ after two decimal places; in other trials after one decimal place to simulate noisy observations. The total number of observations is $2M$, where M is the number of instants at which the populations are observed. The results are given in Table I. The true values of the constants are given at the bottom of the table. The estimated values for each trial are the results of three to five iterations.

TABLE I. Computational Results for the Estimation of Parameters and Initial Conditions of the Population Equations

Observations								
Trial	2M	No. of decimal places	$c_1 = N_1(0)$	$c_2 = N_2(0)$	c_3	c_4	c_5	c_6
1	6	1	1.200	1.000	1.552	0.648	1.669	0.688
2	10	1	1.197	1.014	1.103	0.942	1.146	0.983
3	22	1	1.186	1.019	1.080	0.993	1.119	1.035
4	202	1	1.152	1.058	0.982	1.019	1.026	1.063
5	6	2	1.200	1.090	0.998	1.008	0.999	1.011
6	202	2	1.196	1.096	0.998	1.003	1.002	1.008
True values of constants			1.2	1.1	1.0	1.0	1.0	1.0

Several trends may be noted. First, it is obvious that increasing the accuracy of the data yields superior estimates of parameters. Second, if the accuracy is fixed, increasing the number of observations tends to produce more accurate estimates. Next, looking at the results of trials 5 and 6, we note that six observations (the minimum number required) correct to two decimal places, gives very good results and that 202 observations gives only slightly improved estimates. Then, comparing the constants of trial 4 against those of trial 5, we see that the minimum number of observations correct to two decimals is of greater value in the determination of constants rather than a large quantity of data of lower accuracy.

It is clear that such information, gained through numerical experiments, is of value in planning and interpreting ecological field studies.

BIBLIOGRAPHY AND COMMENTS

Section 1. We are following the paper by R. Bellman, H. Kagiwada and R. Kalaba, "Inverse Problems in Ecology", J. Theor. Biol. 11, 164-167 (1966).

Chapter VI

INDUCED ENZYME SYNTHESIS

1. Introduction

In this chapter we use the quasilinear technique to estimate the initial conditions and the system parameters of a mathematical model for inducible enzyme synthesis. Both quasilinearization and dynamic programming are used to identify the system.

2. The Mathematical Model for Inducible Enzyme Synthesis

In 1964, Heinmets proposed the mathematical model for the process of inducible enzyme synthesis. By way of a brief explanation of the model, we consider the following process.

Plant and animal cells have the ability of "turning on" and "turning off" parts of their metabolic function in accordance with their metabolic needs. This is done by producing (or not producing) those enzymes (catalysts) that are responsible for the manufacture of the metabolic products of importance to the cell. These enzymes whose production can be turned on or off as needed are called inducible enzymes. The production of these enzymes requires the involvement of the cells' hereditary material (DNA) and their mechanisms for protein synthesis.

The bacterial chromosome (DNA) contains all the information for the synthesis of every component of the cell. For each inducible enzyme, there is on the bacterial chromosome a regulator gene X_3 and a functional gene X_7 (see Table I).

TABLE I. Description of Terms used in the Models of Inducible Enzyme Synthesis

Present notation	Description
x_1	Complexing product of regulator gene and metabolic product of the repressing metabolite
x_2	Complexing product of regulator gene and metabolic product of inducer
x_3	Regulator gene
x_4	Repressed functional gene
x_5	An integrated assembly of messenger RNA precursors on the functional gene
x_6	Repressor molecule, synthesized by regulatory gene and $S_r(S_R)$ complex
x_7	Functional gene for messenger RNA synthesis; has a site for interaction with repressor, which leads to repressed functional gene
x_8	Ribosomes
x_9	Messenger RNA specific to functional gene
x_{10}	Newly formed enzyme separated from the template
x_{11}	An integrated assembly of amino acids on the template
x_{12}	Template for synthesis of the enzyme

The functional gene contains the genetic information for the construction of the enzyme. The regulator gene is responsible for the turning on and off of the functional gene. Each of these genes has on it two sites at which the components of the system can react with the gene.

It has been demonstrated experimentally that an inducible enzyme is produced by the bacterial cell only after the cell is grown in the presence of the substrate of the enzyme, that is, the compound that the enzyme digests for the cell. Hence the substrate must get into the cell and somehow come to the attention of the DNA (regulator gene) in order to start the production of the needed enzyme. This also means that in the absence of the substrate the manufacture of enzyme must somehow remain turned off. It was postulated that this machinery remained turned off because a "repressing metabolite" was always present in the cell. This substance undergoes some chemical changes within the cell before becoming activated. It then forms a reversible complex X_1 with one of the sites of the regulator gene. This complex interacts with the cellular pool of precursors for the manufacture of the active repressor, and thus the active repressor X_2 is formed, with the regeneration of the free site on the regulator gene. The repressor then travels to one of the interaction sites on the functional gene and forms a complex with this site, x_1, resulting in a repressed functional gene x_4. The complex is irreversible, but of finite duration, because the repressor is thought to have a finite lifetime. As long as the repressor is complexed with the gene, the latter cannot direct the production of any enzyme, and thus remains repressed. Thus in the absence of the enzyme's substrate, since the cell always contains the repressing metabolite, the repressor is always made in sufficient amounts to keep the site on the functional gene complexed at all times, thereby keeping the enzyme manufacturing machine turned off.

When the cells are grown in the presence of the substrate or "inducer", this enters the cells and may undergo some chemical transformations into an "active" molecule. The inducer was thought to interact in some way with either the formation or function, or both, of the repressor, by acting either (1) at the level of the regulatory gene, by complexing with its second site and forming a complex, x_2; (2) at the level of the repressor-functional gene complex; or (3) directly on the repressor itself. If enough inducers are formed, they can interrupt repressor action, and since the repressor-functional gene complex is of finite duration, eventually the site on the functional gene where the repressor would complex would be left open and the functional gene x would set into action the machinery for the production of the inducible enzyme.

The functional gene, now no longer repressed, reacts at its other site with the pool of RNA components. These attach themselves to the gene in a specific order x_5 to form a molecule of messenger RNA x_7. This specific arrangement of the RNA precursors is a "code" which contains the information for the structure of the enzyme to be manufactured. The messenger RNA detaches itself from the functional gene and travels to the ribosomes x_8 and attaches itself there, forming the template x_{12} for the formation of the enzyme. The ribosomes move along the template "reading off" the series of template code words, and allowing the transferred RNS molecule (with its attached amino acid) which has the code word corresponding to the template code word to attach itself to that template code word. The amino acids are then attached to each other (x_{11}) and the transferred RNA's of the joined amino acids are disconnected from them and the template and allowed to return to the cytoplasm for future use. In this way the chain of amino acids grows, and eventually the last amino acid is put on the chain, and the enzyme (x_{10}) and template (x_{12}) separate. The messenger RNA, template (x_{12}) and ribosomes (x_8) separate, with the possible formation of other disintegration products.

3. Quasilinearization and Differential Approximation

In Sec. 2 we described the heuristic model governing the process of induced enzyme synthesis. The set of equations are given in (3.1).

$$
\begin{aligned}
\dot{x}_1 &= a_1x_3 - a_2x_1 - a_3x_1 \; ; \\
\dot{x}_2 &= a_{15}x_3 - a_4x_2 \; ; \\
\dot{x}_3 &= - a_1x_3 + a_2x_1 + a_3x_1 - a_{15}x_3 + a_4x_2 \; ; \\
\dot{x}_4 &= a_{12}x_7x_6 - a_7x_4 \; ; \\
\dot{x}_5 &= a_6x_7 - a_8x_5 \; ; \\
\dot{x}_6 &= a_3x_1 - a_5x_6 - a_{12}x_6x_7 \; ; \\
\dot{x}_7 &= - a_{12}x_6x_7 + a_7x_4 - a_6x_7 + a_8x_5 \; ; \\
\dot{x}_8 &= a_{14}x_{12} - a_{13}x_8x_9 \; ; \\
\dot{x}_9 &= a_8x_5 - a_9x_9 - a_{13}x_8x_9 \; ; \\
\dot{x}_{10} &= a_{11}x_{11} \; ; \\
\dot{x}_{11} &= a_{10}x_{12} - a_{11}x_{11} \; ; \\
\dot{x}_{12} &= a_{13}x_8x_9 - a_{10}x_{12} + a_{11}x_{11} - a_{14}x_{12} \; .
\end{aligned}
\tag{3.1}
$$

The initial conditions are given as

$$x_i(0) = C_i \; , \qquad i = 1,2,\ldots,12 \; .$$

The system of Eqs. (3.1) forms a complete mathematical problem, for if the coefficients a's and the initial conditions C's are known, the equations can be integrated numerically to obtain a solution.

However our task is quite different, for we are interested in the inverse problem. From this point of view, the coefficients or rate constants and the initial conditions are unknown, and the components $x_i(t)$ are known by experiment. Our problem is to determine the coefficients and initial conditions in such a way that the solution of the system of Eqs. (3.1) will best fit the experimental data. In a sense we are seeking to discover how the system started, the state of the system at time zero, and the interrelationships between the components of the system. In other words, we want to know its structure. The remaining part of this section is devoted to developing the mathematical apparatus for solving the inverse problem.

It is convenient to write the system of Eqs. (3.1) in vector form. We then have

$$\frac{dX}{dt} = AX + B \quad ,$$

$$X(0) = c \quad , \tag{3.2}$$

where $X(t)$ is the unknown vector, A is the matrix of coefficients, and B is a vector containing the nonlinear terms.

Since the constants of the system are also considered as unknowns, it is convenient to treat them in the same manner as the initial conditions.

If we introduce into the system the vector,

$$\alpha = \begin{bmatrix} a_1 \\ a_2 \\ \cdot \\ \cdot \\ \cdot \\ a_{14} \end{bmatrix}$$

then the final system can be written,

$$\begin{aligned} \frac{dX}{dt} &= A(\alpha)X + B(X,\alpha) \\ \frac{d\alpha}{dt} &= 0 \quad , \\ X(0) &= c \quad , \\ \alpha(0) &= \alpha_0 \quad . \end{aligned} \tag{3.3}$$

Through this we have been able to achieve the following. All unknowns of the inverse problem now appear as initial conditions. The penalty we pay is an increase in the size of the system.

Rewriting the vector Eqs. (3.3) in component form, we obtain

$$\frac{dx_i}{dt} = \sum_{j=1}^{12} A_{ij}x_j + b_i \quad ,$$

$$x_i(0) = c_i \ , \qquad i = 1,2,\ldots,12 \quad ,$$

(3.4)
$$\frac{da_k}{dt} = 0 \quad ,$$

$$a_k(0) = (a_k)_0 \ , \qquad k = 1,\ldots 14 \quad .$$

We note that this system has a total of 26 unknowns (12 initial conditions and 15 unknown constants, the latter listed in Table II).

The quasilinear technique allows us to write,

$$\frac{dx_i^{(n+1)}}{dt} = \sum_{m=1}^{12} B_{im}x_m^{(n+1)} + \sum_{k=1}^{14} C_{ik}a_k^{(n+1)} + D_i \ , \qquad i = 1,\ldots 12 \, ,$$

(3.5)
$$\frac{da_k^{(n+1)}}{dt} = 0 \ , \qquad k = 1,\ldots,14 \quad ,$$

where

$$B_{im} = A_{im}(a^{(n)}) + \frac{\partial}{\partial x_m} (b_i(X^{(n)}, a^{(n)}) \quad ,$$

$$C_{ik} = \sum_{j=1}^{12} \frac{\partial}{\partial A_k} A_{ij}(a^{(n)})x_j^{(n)} + \frac{\partial}{\partial a_k} b_i(X^{(n)}, a^{(n)}) \quad ,$$

$$D_i = \sum_{j=1}^{12} A_{ij}(a^{(n)})x_j^{(n)} + b_i(X^{(n)}, a^{(n)})$$

$$- \sum_{l=1}^{12} B_{il}x_l^{(n)} - \sum_{k=1}^{14} C_{ik}a_k^{(n)} \quad .$$

TABLE II. Description of the Rate Constants (Based on Positive Values)

Present notation	Description
a_1	Increase in concentration of the complex product x_1 per unit activity of the regulator gene x_3
a_2	Decrease in concentration of the complex product x_1, per unit concentration
a_3	Rate at which a unit concentration of the complex product x_i will decrease due to its reaction with a constant pool for repressor formation
a_4	Rate at which the unit concentration of the complex product x_2 decomposes
a_5	Rate of decrease of a unit concentration of the repressor molecule x_6 due to decomposition
a_6	Rate of increase of the precursors x_5 due to a unit activity of the functional gene x_7
a_7	Rate of decrease in the activity of the newly formed non-functional repressed gene x_4
a_8	Rate of decrease of a unit concentration of precursors x_5 forming messenger RNA x_9
a_9	Rate of decrease in the concentration in messenger RNA x_9 by dissociation
a_{10}	Rate of increase in the concentration of amino acid template complex x_{11} due to a unit concentration of template x_{12}
a_{11}	Rate of increase in concentration of the enzyme x_{10} due to a unit concentration of the template x_{11}
a_{12}	Rate of increase of activity of the repressed functional gene x_4 due to the combined reaction of a unit concentration of the repressor molecule x_6 and a unit active functional gene x_7

.../

TABLE II. - Cont'd

Present notation	Description
a_{13}	Rate of increase in the concentration of template x_{12} due to the combined reactions of a unit concentration or ribosomes x_6 and a unit concentration of messenger RNA x_9
a_{14}	Rate of increase in the concentration of ribosomes x_8 due to the decomposition of a unit concentration of template x_{12}
a_{15}	External inducer concentration, the rate of increase in the concentration of the complex product x_2 when the activity of the regulator gene x_3 has a value of unity

This is a system of 26 coupled linear differential equations with non-constant coefficients. The cross coupling matrices are $B(12 \times 12)$ and $C(12 \times 14)$ while $D(12)$ forms the forcing function. The structure of this system shows that the last 14 equations are constant in time, a fact which we will use later in the analysis.

4. The Application of the Experimental Data

We recall from the previous section that the solution of the set of quasilinear differential equations can be expressed as

$$X^{(n+1)} = P^{(n+1)} + \sum_{k=1}^{12} C_k^I H_k^{(n+1),I} + \sum_{l=1}^{14} C_l^{II} H_l^{(n+1),II} \tag{4.1}$$

where, using the quasilinear technique, $H_k^{(n+1),I}$ are the homogeneous vectors corresponding to the initial conditions and $H_l^{(n+1),II}$ correspond to the system parameters.

Our problem is to find the set of constants C_k^I and C_l^{II} and we make use of the experimental data to do this.

Let us assume that an experiment has been carried out in the laboratory on the induced synthesis of a particular enzyme, and the time history of all 12 components have been carefully measured over the time interval (0,T). Let these measured components be $\xi_i(t)$, then the error vector $\psi(t)$ can be introduced with the components

$$\psi_i(t) = \xi_i(t) - \chi^{(n+1)}, \quad i = 1,2,\ldots,12 .$$

or

$$(4.2) \qquad \psi_i(t) = \xi_i(t) - \left\{ P_i^{(n+1)} + \sum_{k=1}^{12} C_k^I H_k^{(n+1),I} + \sum_{l=1}^{14} C_l^{II} H_l^{(n+1),II} \right\}.$$

We require that the solution best fit the observed data over the time span (0,T), that is

$$(4.3) \qquad \int_0^T \psi(t) \cdot \psi(t) dt$$

be a minimum.

5. Improving the Mathematical Model

So far we have assumed that the rate constants, although unknown, do not vary over the time interval (0,T). We now consider the situation in which these constants may change over the time interval.

Let us assume that some of the rate constants do change in the interval (0,T), dt, with changes occurring at specific, but unknown, times within the interval. We define as critical times those in which one or more of the rate constants change values. The problem facing us now is to find the critical times together with the initial conditions at each critical time, and the rate constants over each critical interval. If this information can be found, an improved model of the system can be established whose proper interpretation may lead to a better understanding of the physical mechanism in operation.

In the last section we showed how a mathematical model can be made to fit the observed data. In a similar manner, if the time interval is divided into subintervals, we can associate with each subinterval an error in the data fit. If the rate constants change during the interval $(0,T)$, then the entire error in the data fit would be smallest only if the data fit is made over each correct subinterval. The entire process of finding the correct subintervals, together with the appropriate rate constants and initial conditions at the beginning of each subinterval, is called segmental differential approximation.

To introduce this new idea we consider first the problem of minimizing the linear function

$$E(x_1,x_2,\ldots,x_M) = e_1(x_1) + e_2(x_2) +\ldots+ e_M(x_M) \quad ,$$

(5.1) $$e_i(x_i) \geq 0 \ , \qquad x_i \geq 0 \ , \qquad \sum_{i=1}^{M} x_i = x \quad .$$

The problem (5.1) can be considered to be imbedded in a class of problems in which M is any positive integer and x any positive value. We view the process as dynamic by requiring the choice of x_i to be made one at a time.

Since E must depend on the region x and the number of subregions M, let us define a function

$$F_M(x) = \min_{x} E(x_1,x_2,\ldots,x_M) \quad ,$$

$$\sum_{i=1}^{M} x_i = x \ , \qquad x_i \geq 0 \quad .$$

It is clear that

$$F_M(0) = 0 \ , \qquad M = 1,2,\ldots,$$

if

$$e_i(0) = 0 \quad \text{for all} \quad i \ ,$$

and

$$F_1(x) = e_1(x) \ , \qquad x \geq 0 \quad .$$

To obtain a recurrence relationship between $F_M(x)$ we proceed as follows: select $x_M (0 \le x_M \le x)$. Then, regardless of the value of x_M, there are $M-1$ remaining choices in the region $(x - x_M)$. Then for a given x_M the minimum E is

$$e_M(x_M) + F_{M-1}(x - x_M) \quad .$$

Since x_M is arbitrary, then clearly,

$$F_M(x) = \min_{\{0 \le x_M \le x\}} [e_M(x_M) + F_{M-1}(x - x_M)] \quad ,$$

$$F_1(x) = e_1(x) \quad , \tag{5.2}$$

which is the fundamental equation of dynamic programming.

To make the previous analysis applicable to the enzyme problem, let the time interval of observation between 0 and T be subdivided by a sequence of intermediate times $\{t_k\}$. We assume that the critical times form a small subset of $\{t_k\}$, namely $\{t_k\}_i$. Thus we find the correct interval by applying the technique of dynamic programming.

BIBLIOGRAPHY AND COMMENTS

1. Section 1. We are following the paper by R.S. Roth and M.M. Roth, "Data Unscrambling and the Analysis of Inducible Enzyme Synthesis", Math. Biosci., 5, 57-92 (1969).

2. Additional references are

   i. F. Heinmets, "Analog Computer Analysis of a Model System for the Induced Enzyme Synthesis", J. Theor. Bio., 6, 60-75 (1964).

   ii. A.B. Pardee and L.S. Pretidge, Biochem. Biophys. Acta, 49, 77 (1961).

   iii. R.E. Bellman, B. Gluss and R.S. Roth, Segmental Differential Approximation and the Black Box Problem, RM-4269-PR, (RAND Corporation, October 1964).

iv. R.E. Bellman and R.E. Kalaba, *Quasilinearization and Nonlinear Boundary Value Problems*, (American Elsevier, New York, 1965).

v. R.E. Bellman, H. Kagiwada and R.E. Kalaba, *Quasilinearization, System Identification, and Prediction*, RM-3812-PR, (RAND Corporation, August 1963).

vi. R.E. Bellman, *Dynamic Programming*, (Princeton Univ. Press, Princeton, New Jersey, 1957).

vii. R.E. Bellman and S.E. Dreyfus, *Applied Dynamic Programming*, (Princeton Univ. Press, Princeton, New Jersey, 1962).

viii. J.D. Watson, *Molecular Biology of the Gene*, (Benjamin, New York, 1965).

ix. F. Jacob and J. Monod, "Genetic Regulatory Mechanisms in the Synthesis of Proteins", *J. Mol. Bio.*, *3*, 318-356 (1961).

x. S.H. Crandall, *Engineering Analysis: A Survey of Numerical Procedures*, (McGraw-Hill, New York, 1956).

xi. R.E. Bellman, R.E. Kalaba and J.A. Lockett, *Numerical Inversion of the Laplace Transform*, (American Elsevier, New York, 1966).

xii. R.S. Roth, "The Unscrambling of Data: Studies in Segmental Differential Approximation", *J. Math. Anal. Appl.*, *14*, 5-22 (1966).

xiii. R.E. Bellman, B. Gluss and R.S. Roth, "On the Identification of Systems and the Unscrambling of Data: Some Problems Suggested by Neurophysiology", *Proc. Nat. Acad. Sci.*, USA, *52*, 1239-1240 (1964).

xiv. R.E. Bellman and R.S. Roth, "A Technique for the Analysis of a Broad Class of Biological Systems", Bionics Symp. 1966, in *Cybernetic Problems in Bionics*, (Gordon and Breach, New York and London, pp. 725-737).

xv. R.E. Bellman, J. Jacquez, R. Kalaba, and S. Schwimmer, "Quasilinearization and the Estimation of Chemical Rate Constants from Raw Kinetic Data", *Math. Biosci.*, *1*, 71-76 (1967).

3. For differential approximation, refer to the book by R. Bellman, *Methods of Nonlinear Analysis*, Vol. I, (Academic Press, Inc., New York, 1969).

4. For the theory of dynamic programming, see

   i. S. Dreyfus and A.M. Law, *The Art and Theory of Dynamic Programming* (1977).

   ii. R. Bellman, *Introduction to the Mathematical Theory of Control Processes*, Vol. I, (Academic Press, Inc., 1969).

   iii. R. Bellman, *Introduction to the Mathematical Theory of Control Processes*, Vol. II, (Academic Press, Inc., 1971).

Chapter VII

DISTRIBUTED SYSTEMS

1. Introduction

In recent years, a considerable degree of experience has been gained in the numerical solution of various system identification problems. This is because modern digital computers can be very effective in the numerical solution of nonlinear multi-point boundary-value problems for large systems of ordinary differential equations. Generally speaking, when studying systems governed by more complex functional equations, the approach has been to approximate the original equation by a system of ordinary differential equations. This approach has worked well with various inverse problems in radiative transfer, where the underlying equations are differential-integral equations. It has also been employed in system identification problems involving differential-difference equations. In this chapter, we wish to indicate some of the approaches open to us in the identification of distributed systems described by partial differential equations.

2. The Telegrapher's Equation

Consider a propagation process taking place in a one-dimensional medium of length L. Let the describing equation be the telegrapher's equation

$$(2.1)\qquad \frac{\partial^2}{\partial x^2} u = a \frac{\partial^2 u}{\partial t^2} + b \frac{\partial u}{\partial t} + cu \quad ,$$

$$(2.2)\qquad 0 \leq x \leq L \quad , \qquad t > 0 \quad ,$$

where a, b and c are certain system constants to be estimated on the basis of observations over time. The medium is at rest initially,

$$(2.3)\qquad u(x,0) = 0 \ , \qquad u_t(x,0) = 0 \ , \qquad 0 < x < L \quad ,$$

and the left-hand end is held fixed,

$$(2.4)\qquad u(0,t) = 0 \quad .$$

The input to the system at the right-hand end, $x = L$ is given by

$$(2.5)\qquad u_x(L,t) = f(t) \ , \qquad t > 0 \quad .$$

Suppose that the resulting observed response of the system at $x = L$ is

$$(2.6)\qquad u(L,t) \cong g(t), \qquad 0 < t \quad .$$

It is indicated that $u(L,t)$ is approximately $g(t)$, since the observations are to be considered noisy, and Eq. (2.1) is not a perfect description of the process.

It is desired to determine the system parameters a, b and c in such a way as to explain the observed response in some prescribed optimal fashion. Now consider the role of Laplace transforms and finite-difference methods in the precise formulation and numerical solution of system identification problems of this nature.

3. Laplace Transforms

We introduce the Laplace transforms

$$(3.1)\qquad L(u) = U(x,s) = \int_0^\infty e^{-st}\, u(x,t)dt \quad ,$$

(3.2) $$L(f) = F , \qquad L(g) = G ,$$

and find

(3.3) $$U'' = (as^2 + bs + c)U ,$$

where the prime indicates differentiation with respect to x, and s is regarded as a parameter. As auxiliary conditions, we have

(3.4) $$U(0,s) = 0$$

and

(3.5) $$U_x(L,s) = F(s) .$$

In addition, from the observations, we have

(3.6) $$L(u(L,t)) = L(g(t)) = G(s) .$$

We assume that $G(s)$ is evaluated for various values of s, either real or complex, by numerical integration, giving

(3.7) $$G(s_j) = b_j , \qquad j = 1,2,\ldots,M .$$

We now select the constants a, b and c so as to minimize the sum S, where

(3.8) $$S = \sum_{j=1}^{N} \{U(L,s_j) - b_j\}^2 .$$

The value of $U(L,s_j)$ is determined by solving Eq. (3.3), subject to the two-point constraints in Eqs. (3.4) and (3.5). This minimization can be done via quasilinearization, and some successful experiments for the wave equation are described below.

By their nature, a, b and c are non-negative. If positive values of s are chosen then Eq. (3.3) is unstable, which is undesirable numerically. To avoid this difficulty, observe that we can express $U(L,s)$ in the form

(3.9) $\quad U(L,s) = F(s)R(L,s)$,

where R is the solution of the stable initial value problem,

(3.10) $\quad \frac{dR}{dL} = 1 - (as^2 + bs + c)R^2 \ , \qquad R(0,s) = 0$.

4. Example

Consider the wave equation,

(4.1) $\quad \frac{\partial^2}{\partial x^2} = \frac{1}{c^2(x)} \frac{\partial^2}{\partial x^2}$,

(4.2) $\quad 0 < x < 1 \ , \qquad 0 < t$.

The speed c is a function of position,

(4.3) $\quad c(x) = a_0 + a_1 x$,

where a_0 and a_1 are certain constants to be estimated. The initial and boundary conditions are

(4.4) $\quad u(x,0) = 0 \ , \qquad \frac{\partial}{\partial t}(x,0) = 0 \ , \qquad 0 < x < 1$,

(4.5) $\quad u(0,t) = 0 \ , \qquad \frac{\partial}{\partial x}(1,t) = \delta(t) \ , \qquad 0 < t$,

where $\delta(t)$ is the Dirac delta function. The given observations are

(4.6) $\quad u(1,t) \cong g(t) \ , \qquad t > 0$.

The constants a_0 and a_1 in Eq. (4.3) are to be estimated on the basis of the measurements, using the scheme involving Laplace transforms.

The Laplace transform of the observations,

(4.7) $\quad U(1,s) = L(u(1,t)) \cong b_s \ , \qquad s = 1,2,\ldots,7$, is given in Table I.

TABLE I. Values of the Laplace Transform

s	b_s
1	0.811967
2	0.551174
3	0.390695
4	0.297835
5	0.239837
6	0.200613
7	0.172392

The quasilinearization method is used to determine the parameter a_0 and a_1, which make the sum

(4.8) $$S = \sum_{s=1}^{7} \{U(1,s) - b_s\}^2$$

a minimum. As initial estimates,

(4.9) $$a_0 \cong 1.2 , \qquad a_1 \cong 0.3 .$$

At convergence,

(4.10) $$a_0 = 0.9996 , \qquad a_1 = 0.50005 .$$

The correct values are

(4.11) $$a_0 = 1.0 , \qquad a_1 = 0.5 .$$

Now we also know the missing "initial" conditions

(4.12) $$U_x(0,s) , \qquad s = 1,2,\ldots,7 .$$

If steady-state measurements of the reflected wave are given, the estimation of constants may then be based on the Riccati equation for the reflected wave.

5. Semi-discretization

To approximate $\frac{\partial^2}{\partial x^2}(x_i,t)$, write

$$(5.1)\qquad \frac{\partial^2}{\partial x^2}(x_i,t) \cong \sum_{j=1}^{N} b_{ij}\, u(x_j,t) \quad .$$

The well-known centered second difference expressions can be used. The approximate system of ordinary differential equations for the telegrapher's equation can be written

$$(5.2)\qquad u(x_i,t) = u_i(t) \ , \qquad i = 1,2,\ldots,N \ ,$$

$$(5.3)\qquad a\ddot{u}_i + b\dot{u}_i + cu_i = \sum_{j=1}^{N} b_{ij}u_j \ , \qquad i = 1,2,\ldots,N \ .$$

If it is assumed that the solution of the system of ordinary differential equations in Eq. (5.1) yields a good approximation of the solution to the telegrapher's equation, then it is again necessary to estimate the constants appearing in a system of ordinary differential equations.

6. Finite Difference Methods

Let us consider a semi-infinite inhomogeneous elastic rod extending from $0 \leq x < \infty$. The rod is initially at rest and a stress is applied to the left-hand end, $x = 0$. The resulting time history of the velocity of the end is observed. We wish to deduce the local wave speed as a function of position. The basic equations along the characteristics are

$$(6.1)\qquad \frac{\partial u'}{\partial \alpha} - \rho c u_\alpha^2 = 0 \ ,$$

(6.2) $$\frac{\partial u'}{\partial \beta} + \rho c u^2_\beta = 0 , \qquad x > 0 , \qquad t > 0 ,$$

with

(6.3) $$u^1(x,0) = 0 ,$$

(6.4) $$u^2(x,0) = 0 , \qquad x > 0 .$$

The characteristic curves are

(6.5) $$\frac{dx}{dt} = \pm c(x) ,$$

where $u^1 = \rho c^2 \frac{\partial u}{\partial x}$ is the stress, $u^2 = \frac{\partial u}{\partial t}$ and α and β are the coordinates along the positive and negative characteristics, respectively. Assume ρ, the density, equals unity.

The observations

(6.6) $$g(t_i) \cong u^1(0,t_i) ,$$

(6.7) $$m(t_i) \cong u^2(0,t_i) ,$$

are given at discrete times, t_i. We wish to estimate $c(x)$ at discrete points, x_i.

Consider a region R bounded by the t-axis, the negative characteristic through $x = 0$, $t = t_n$, and the positive characteristic through $x = 0$, $t = 0$. Suppose $u^1(x,t)$, $u^2(x,t)$ and $c(x)$ have been estimated on a suitable grid of characteristics in R and on its boundary. The observations at $x = 0$, $t = t_{n+1}$, the prior estimates along the negative characteristic through $x = 0$, $t = t_n$, and a finite difference approximation to Eqs. (6.1) and (6.2) may be used to estimate the solution at various points along the negative characteristic given $x = 0$, $t = t_{n+1}$. This process may be continued until the negative characteristic through $x = 0$, $t = t_{n+1}$ intersects the positive characteristic through the origin. Let P denote this point. Let A and B denote the previous points along the positive and negative characteristics through P, respectively. Denote u^1 at $x = x_P$, $t - t_P$ by u^1_P, and similarly for the other variables (see Fig. 7.1).

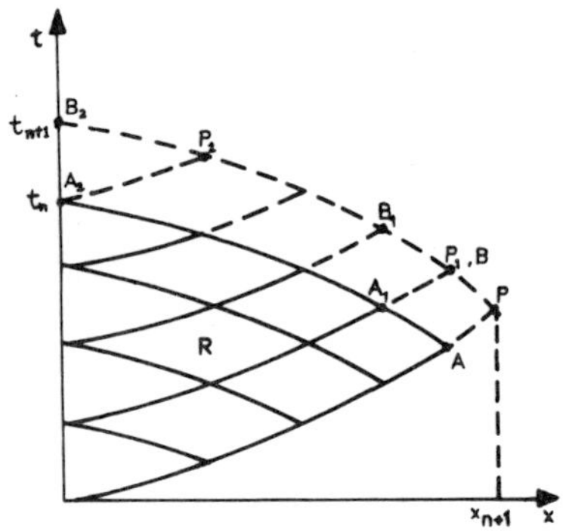

Fig. 7.1 The Net of Characteristic Curves

At P, u^1_P, u^2_P and c_P are unknown. A finite difference approximation to Eqs. (6.1) and (6.2) at P is

$$(6.8) \qquad (u^1_P - u^1_A) - \tfrac{1}{2}(c_P + c_A)(u^2_P - u^2_A) = 0 \quad ,$$

$$(6.9) \qquad (u^1_P - u^1_B) + \tfrac{1}{2}(c_P + c_B)(u^2_P - u^2_B) = 0 \quad .$$

For the region $t < t_P$ and $x > x_P$, u^1 and u^2 are both zero. If, for example, a discontinuity is propagating along the positive characteristic through the origin, then Eq. (6.2) at P becomes

$$(6.10) \qquad u^1_P + c_P u^2_P = 0 \quad .$$

Equations (6.8), (6.9) and (6.10) may be solved to yield u^1_P, u^2_P and c_P. Equation (6.10) may be modified if the solution is continuous across the positive characteristic through the origin. Then the wave speed, c, at the new position, x_P, may be estimated. The error in the estimated c may be related to the error in u^1_P and u^2_P.

Some numerical experiments have shown the efficacy of the method suggested. The essential features of the wave speed function have been determined using quite noisy observations.

7. Example

Some controlled numerical experiments were performed to study the solutions to the direct and inverse problems. The experiments were divided into three phases. In Phase I the solution to the direct problem was computed. In Phase II the data from Phase I was corrupted with random noise. Finally, in Phase III the noise-corrupted data were used in the inverse problem. In this fashion, precise control was maintained over the data used in the inverse problem.

The model used in Phase II for the simulated measurements was

$$y_i = x_i + P_1 \cdot r_i^1 + P_2 \cdot r_i^2 \cdot |x_i| \quad , \tag{7.1}$$

where

x_i = the variable measured,

y_i = the measured value of x_i,

r_i^1, r_i^2 = statistically independent Gaussian random numbers with zero mean and unit variance,

(7.2) P_1, P_2 = parameters to adjust the level of the noise.

The first noise term in Eq. (7.1) masks out low level signals and zero crossings. The second noise term in Eq. (7.1) controls the average percentage error.

Figures 7.2 and 7.3 display the results for the direct problem. The local wave speed $c(x)$ and input pulse $u^1(0,t)$ are shown in Figs. 7.2 and 7.3. The resulting response $u^2(0,t)$ is given in Fig. 7.3. The mesh size in the t direction was 0.02.

The $u^1(0,t)$ and $u^2(0,t)$ data from Phase I was then used in Phase II. The parameters for the noise model of Eq. (7.1) were as

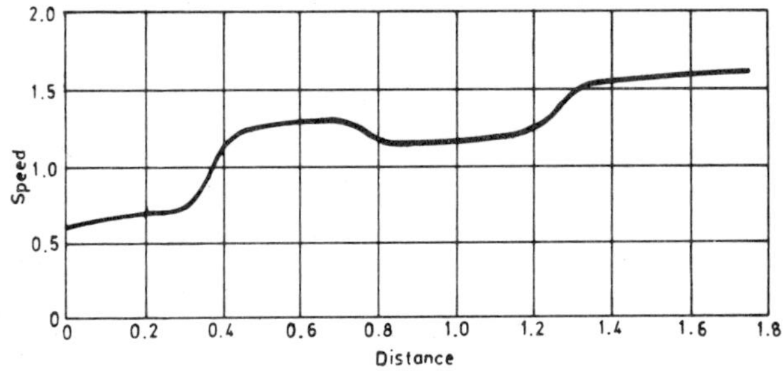

Fig. 7.2 Local Wave Speed vs Distance from Free End

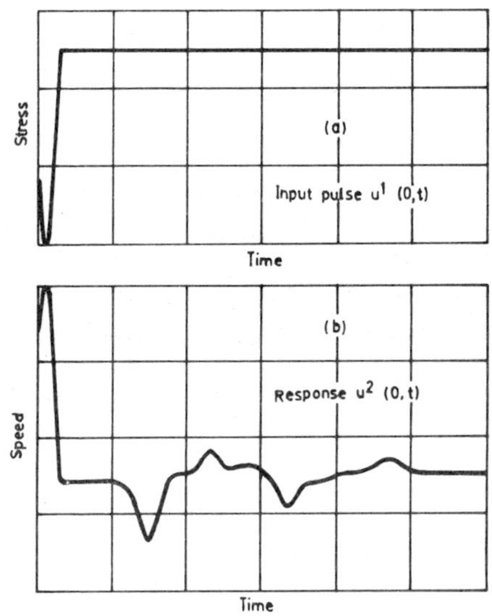

Fig. 7.3 The Input and Response Curves

shown in the table.

Run	$u^1(0,t)$		$u^2(0,t)$	
	P_1	P_2	P_3	P_4
1	0.0	0.0	0.0	0.0
2	0.0	0.01	0.0	0.01
3	0.001	0.01	0.001	0.01
4	0.0	0.05	0.0	0.05

The results were used in Phase III. Figure 7.4 displays the estimated $c(x)$ for the 4 runs. (N.B. Each successive $c(x)$ is displayed upward.)

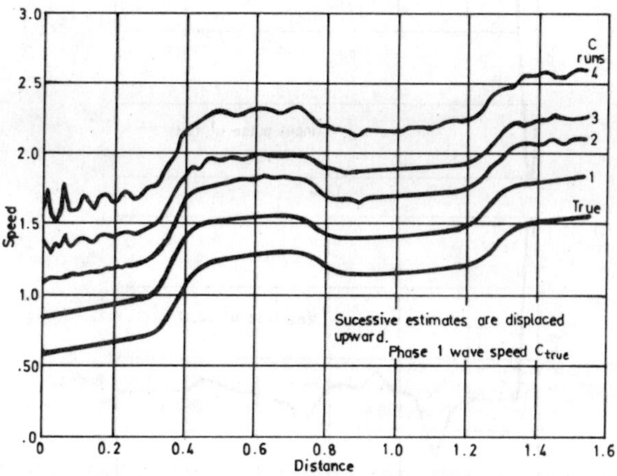

Fig. 7.4 Results for the Inverse Problem

Other numerical experiments have been performed using more complex input pulses and more complex $c(x)$'s with additional and sharper transitions. The results from these experiments were essentially the same as those presented.

8. Discussion

The results from the numerical experiments are encouraging. They should be viewed as the first primitive attempt to attack inverse problems in wave propagation directly. The door has now been opened to many interesting and significant problems.

BIBLIOGRAPHY AND COMMENTS

Section 1. We are following the paper by R. Bellman, D. Detchmendy, H. Kagiwada, and R. Kalaba, "On the Identification of Systems and the Unscrambling of Data III: One-Dimensional Wave and Diffusion Processes", *J. Math. Anal. Appl.*, 23, No. 1, 173-182 (July 1968).

Chapter VIII

HIDDEN PERIODICITIES

1. Introduction

The greater part of classical and contemporary mathematical analysis is devoted to the problem of predicting the future behavior of a system, given the defining equations and the initial state. This is rather surprising since the basic scientific problem, in such fields as physics, engineering, biology, medicine, and economics, to mention only a few, is that of determining the structure of the system, given various observations over time. Actually, the problem is a deeper one of determining the best approximate system that agrees sufficiently well with the available data. Furthermore, what is "best" involves a consideration of the intended use of the model.

The identifications of "inverse" problems are of a high degree of mathematical complexity. The numerical aspects are, however, of an even higher degree of complexity. Due to the emergence of the fast computer with a reasonably large rapid access storage ("memory"), some of these inverse problems have become tractable.

2. An Unscrambling Problem

A basic fact of experimental life is that measuring devices

"hear" many different signals from many different sources. A fundamental problem, then, is that of determining what source provided what stimulus. The classical periodogram analysis problem is of this genre. Given the values $f(t_i)$ where $f(t) = \sum_{i=1}^{N} \alpha_i \cos w_i t$, we wish to determine the parameters α_i and w_i. Another problem of frequent occurrence is that of determining the λ_i's, given the values $f(t_i)$, where $f(t)$ now has the form $f(t) = \sum_{i=1}^{N} \beta_i e^{\lambda_i t}$. These problems are notoriously difficult.

As we have pointed out before, these problems are special cases of determining the coefficients a_i and the initial values c_i, when we know that $f(t)$ satisfies the linear differential equation $f^{(N)} = \sum_{i=0}^{N-1} a_i f^{(i)}$, $f^{(i)}(0) = c_i$, $i = 0,1,2,\ldots,N-1$. This, in turn, is a special case of the problem of determining the vectors a^i and the initial vectors c^i, when $f(t)$ is known to have the form $f(t) = \sum_{i=1}^{N-1} u_i(t,a^i)$, where the functions $u_i(t,a^i)$ are solutions of nonlinear differential equations of the form $\frac{du_i}{dt} = g_i(u,a^i)$, $u_i(0) = c^i$. We call this the problem of "differential approximation". Particular versions of this problem, in the case where the defining equations are linear, have been treated by Medgyessy, using quite a different approach from the one which we shall employ. References to biomedical and biochemical applications may be found there.

3. The Approach

Suppose that the N-dimensional vector $X(t)$ is a solution of the differential equation $\frac{dX}{dt} = F(X,A)$, $X(0) = C$, where the system parameter vector A and the initial state vector C are as yet unspecified. At certain times the system is "observed", which results in the approximate conditions $(\beta_i, X(t_i)) \cong b_i$, $i = 1,2,\ldots,M \geq N$. We wish to determine the vectors A and C so that we minimize S, the sum of the squares of the deviations $S = \sum_{i=1}^{M} \{(\beta_i, X(t_i)) - b_i\}^2$. If we consider the vector A to be a variable, but subject to the condition $\frac{dA}{dr} = 0$, then we can adjoin this equation to the original equation. Now our problem is to determine the initial conditions C and $H(0)$ so as to minimize S.

4. Computational Formalism

We wish to minimize a sum $S = \sum_{i=1}^{N} \{(\beta_i, X(t_i)) - b_i\}^2$, where $\frac{d}{dX} = f(X)$, and the minimization is over the initial vector C, $X(0) = C$.

Using the numerical techniques of quasilinearization given in Chapter II, we can determine the optimal initial condition C.

Let us call attention to the fact that we are assuming that the particular and homogeneous solutions can be computed accurately on the interval $0 \le t \le t_i$ and that the indicated system of linear algebraic equations can be solved computationally. In some instances these assumptions are not met, though we have experienced little difficulty in the examples which we have treated. Furthermore, we are assuming that the initial approximation is sufficiently good so that the sequence of approximations converges. When convergent, the rate is very fast, since the algorithm is quadratic in nature; we approximately double the number of correct digits with each additional step.

BIBLIOGRAPHY AND COMMENTS

1. Section 1. We are following the paper by R.E. Bellman, H.H. Kagiwada and R.E. Kalaba, "On the Identification of Systems and the Unscrambling of Data I: Hidden Periodicities", Proc. Nat. Aca. Sci., 53, No. 5, 907-910 (May 1965).

2. The book by P. Medgyessy is Decomposition of Superpositions of Distribution Functions, (Budapest, Alcadémiai Kiadó, 1961).

Chapter IX

Measurement of Heart Parameters

1. Introduction

In this chapter we will study the characteristics of the heart using the data derived from observations of skin potentials obtained in electrocardiograms and vectorcardiograms. These problems have great importance for both diagnosis and therapy. Following orthodox statistical methodology, one would compare the 'normal' ECG curves with those of the patient and thus arrive at clinical conclusions.

An alternate method is to postulate a mathematical model of the heart and compare results of the model equations with the observed data, as is done in many other attempts to understand physical phenomena. In this chapter as in Chapter VI we will take time to describe the mathematical model.

The observed skin potentials are accounted for by the ventricular depolarization of the heart. The modeling was given a great impetus by the pioneer work of Scher and Young who mapped out the sequence of myocardial depolarization in dogs. This was extrapolated to the case of the human heart by Durrer and co-workers. This in turn, was followed by the digital computer simulation of the vector and electrocardiograms using anatomically and physiologically based models of the electrical fields associated with heart activity, work carried out by Sylvester et al. These simulations are an attempt to explore the possibility of establishing precise criteria for the diagnosis of varying degrees of ventricular hypertrophies and myocardial infarctions. These simulations proved adequate to throw much light on the understanding of the vectorcardiograms. The further advantage of such simulation is that experiments can be run mathematically and key experiments validated by observations on animals and humans. Since successful simulation validates the model, we need not emphasize the importance of the inverse problem of parameter identification in clinical diagnosis. Bellman *et al.* addressed themselves to this task of identifying the parameters employing the method of quasilinearization, developed for this specific problem. This technique yields excellent estimates provided one starts the iteration procedure with a good initial estimate of some of these unknown parameters. The convergence of course is quadratic and the resulting large systems of equations can be handled by the contemporary digital computer with even greater capabilities in the near future. However, it would be highly advantageous if we could dispense with any approximate *a priori* knowledge of these parameters. The aim of this chapter is to outline such a method where the determination of the essential parameters does not depend on some initial guess. We employ the methods of differential quadrature, aided and abetted by the techniques of splines, to handle the inverse problem. It has turned out that even a small computer can do the computation.

The model will be described in Sec. 2. As a precursor to applying it to an actual situation, we will apply it to the data used in the simulation models and recover the estimates of the parameters used in the above. The method can handle far more sophisticated models of the electrical activities of the heart.

2. The Model

In the model assumed for the simulation purposes the heart is pictured as consisting of 20 segments, 7 in the septal region, 9 in the left ventricle and 4 in the right ventricle. The electromotive surface of the heart is replaced by twenty current dipoles placed at the centroid of these segments of the myocardium (see Fig. 9.1). Sylvester et al. assumed that the activity of each segment can be lumped into a simple current dipole distributed in space, fixed in direction, varying in magnitude as a function of time. In the simulation the sequence of human myocardial depolarization was assumed to be that reported by Durrer. The duration under consideration corresponded to normal human cardiac electrical activity known as the QRS waves of the ECG. This is about 80 msec.

Assuming the body to be an infinite homogeneous conductor in the simulation, the set of current dipoles were placed in a uniform resistive medium and four observation points were placed in the medium at appropriate distances to represent a cubic system of electrode placements on a normal male thorax. In the simulation which is the forward problem, given the time histories of the dipole moments and the potentials measured at the body surface observation points taking one of them to be the reference point, the horizontal, frontal and right sagittal vectorcardiograms can be generated.

Simulated cardiograms resembled to a great extent the clinical cardiograms both in the normal and other than normal situations, as is noted in the references cited above. Figures 9.1 and 9.2 represent some of the features of the model described above.

3. The Inverse Problem and Equations for the Model

The simulation which reproduces the cardiograms represents the forward problem; determining the parameters of the model, namely the h_i, k_i and t_i of the segmental dipoles from the observation of the skin potentials and their time histories, constitutes the inverse problem. This was treated as a multi-point boundary value problem. Starting from an initial guess the parameters were computed employing quasilinearization technique. The disadvantage was that the initial guess had to be sufficiently close to the actual values for the success of the calculation. Here we will detail a method which does not need an

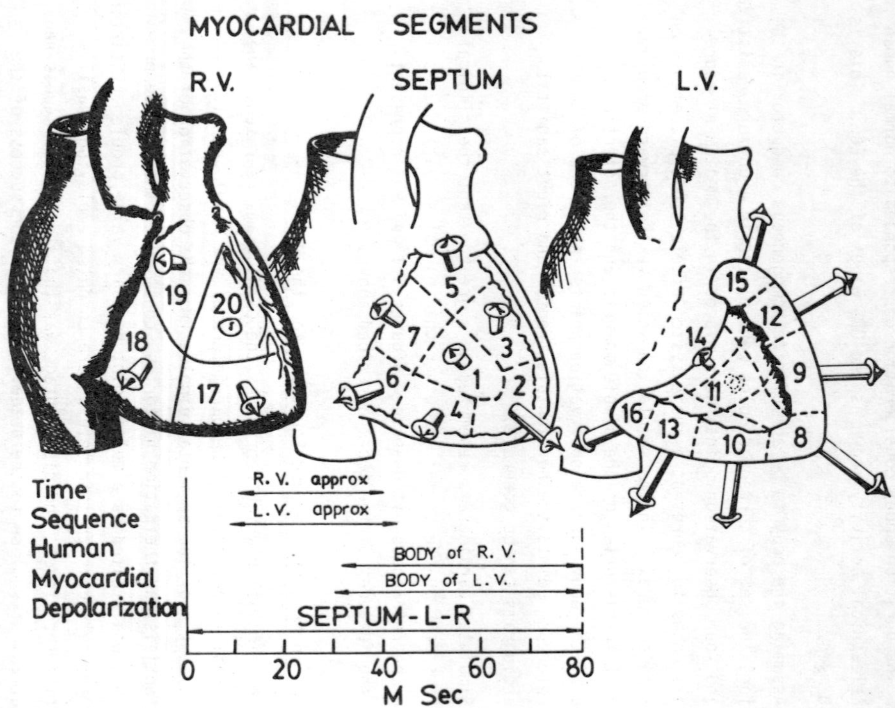

Fig. 9.1 Human Heart Model

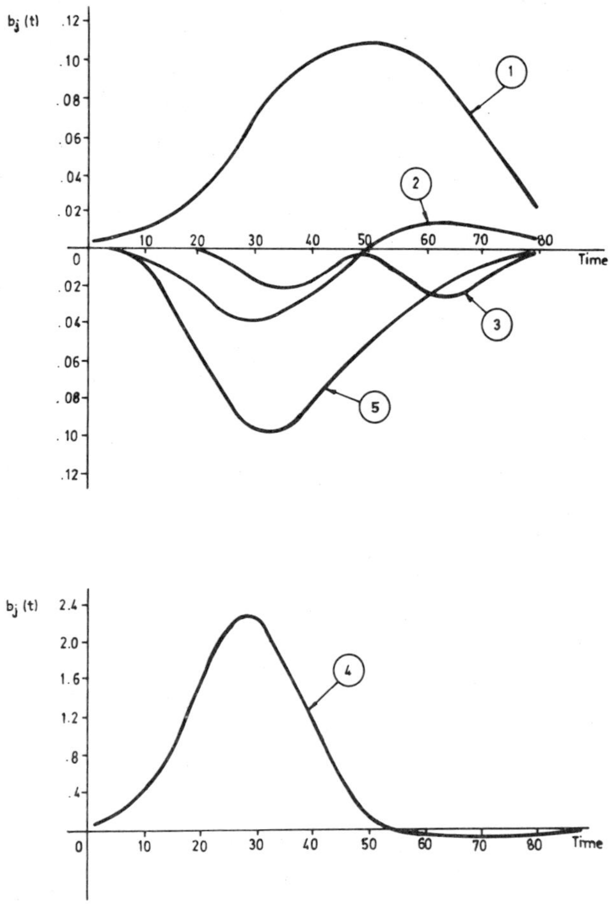

Fig. 9.2 Skin Potentials at the Observation Points for M = 5.

initial guess for their parameters and therefore does not depend on it either.

According to the model, the dipole moment for the i-th segment $(i = 1,2,\ldots,M)$ at time t is represented by $F_i(t)$. This dipole produces a potential $V_j(t)$ at the j-th observation point $(j = 1,2,\ldots,N)$. Equations governing computation of these skin potentials, assuming the human body as a homogeneous resistive medium, are taken to be those described by Wilson and Bailey. Let us introduce the orthogonal axes X_1, X_2 and X_3, then V_j which is the potential at time t at the observation point j is given by

$$(3.1) \qquad V_j(t) = \alpha \sum_i F_i(t) \frac{\cos \theta_{ij}}{|r_{ij}^2|}, \qquad i = 1,2,\ldots,M, \qquad j = 1,2,\ldots,N$$

where θ_{ij} is the angle between the axis of the i-th dipole and line r_{ij} joining the i-th dipole to the j-th observation point. The parameter α is a constant and may be set equal to one. When potential V_j is measured with respect to the potential at a reference point r_{io}, we have

$$(3.2) \qquad V_j(t) = \alpha \sum_i F_i(t) \frac{\cos \theta_{ij}}{|r_{ij}^2|} - \alpha \sum_i F_i(t) \frac{\cos \theta_{io}}{|r_{io}^2|}$$

where $\cos \theta_{io}$ and r_{io} are quantities associated with the reference point $j = 0$. If the reference point is taken at infinite distance the reference potential is zero.

With the activation sequence assumed for the dipoles, $F_i(t)$ is supposed to be described by the following equations:

$$\frac{dF_i(t)}{dt} = k_i(t_i - t)F_i, \qquad 0 \le t \le 80 \text{ msec},$$

$$(3.3) \qquad F_i(0) = h_i, \qquad \frac{dt_i}{dt} = 0, \qquad \frac{dk_i}{dt} = 0, \qquad i = 1,2,\ldots,M .$$

It is assumed that the excitation process is completed in 80 msec and observations of the skin potentials are taken every millisec. As seen from (3.3), t_i is the time at which maximum dipole strength occurs at each segment i, k_i is related to the broadness of the Gaussian curve $F_i(t)$ and h_i is the maximum strength of the dipole at the segment i.

The actual potentials as observed on the surface of the skin at the j-th observation point are given by $b_j(t_\ell)$; $\ell = 1,2,3,\ldots,80$. We wish to determine the dipole functions $F_i(t_\ell)$ at each segment i corresponding to each observation time t_ℓ such that they are in best agreement with the observed potentials $b_j(t_\ell)$. To this end, we minimize the quantity S which is the sum of the squares of the differences of the computed potentials $V_j(t_\ell)$ and the observed quantities $b_j(t_\ell)$ i.e.

$$(3.4) \qquad S = \sum_\ell \sum_j [V_j(t_\ell) - b_j(t_\ell)]^2 \quad .$$

In view of Eq. (3.2) this reduces to

$$(3.5) \qquad S = \sum_\ell \sum_j \left[\sum_i \alpha \left(\frac{\cos \theta_{ij}}{|r_{ij}^2|} - \frac{\cos \theta_{io}}{|r_{io}^2|} \right) F_i(t_\ell) - b_j(t_\ell) \right]^2$$

which is equivalent to minimization of the scalar S given by

$$(3.6) \qquad S = \sum_\ell < AF(t_\ell) - b(t_\ell), \ AF(t_\ell) - b(t_\ell) >$$

where A is an N by M matrix, F an M-dimensional vector and b an N-dimensional vector. When minimized with respect to $F_i(t_\ell)$ we arrive at

$$(3.7) \qquad A^T A F(t_\ell) = A^T b(t_\ell) \ , \qquad \ell = 1,2,\ldots,80 \quad .$$

At this point we want to observe that if $N = M$, (i.e. the number of observation points is the same as the number of dipoles in question), we can easily solve the above system of linear equations and arrive at the values of $F(t_\ell)$ vector at each time instant t_ℓ. For

$N \neq M$, the matrix $A'A$ may be rank deficient and its inversion not possible. Hence we shall only consider the case $N = M$. This is not actually a disadvantage, since it is very easy experimentally to increase the number of observation points to be the same as the number of myocardial segments chosen in the beginning. The maximum number of segments usually considered is 20, so that the maximum dimension for the matrix to be inverted is also 20. This can be numerically solved on any small computer whereas in quasilinearization method, one is forced to use large computers with huge storage locations. Another advantage of this procedure is that it can be done on a real-time basis.

Since each dipole function $F_i(t)$ is numerically known at all time points t_ℓ, we can use interpolation to approximate this function. Cubic spline interpolation approximates the function by a cubic in each time interval and reproduces the values of the function at the grid points. We also have continuity of the function and its first two derivatives at each grid point and any end condition can be easily incorporated.

Defining a set of cardinal splines as spline functions with unit values at the mesh points, i.e.

$$C_\ell(t_k) = \delta_{\ell k} \quad , \tag{3.8}$$

we can approximate $F_i(t)$ as

$$f_i(t) = \sum_\ell C_\ell(t) F_i(t_\ell) \tag{3.9}$$

at each segment $i = 1,2,\ldots,N$. Numerically it is always possible to find a complete set of cardinal splines for the above approximation. Therefore, values of $f_i(t)$ and its first derivative

$$f_i'(t) = \sum_\ell C_\ell'(t) F_i(t_\ell) \tag{3.10}$$

can be calculated at each segment $i = 1,2,\ldots,N$. Substituting these values in Eq. (3.3) reduces the set of ordinary differential equations to a system of linear equations of the form

$$(3.11) \qquad \sum_{\ell}[k_i(t_i - t_\ell)C_i(t_\ell) - C_i'(t_\ell)]\, F_i(t_\ell) = 0\ , \qquad i = 1,2,\ldots,N,\quad \ell = 1,2,\ldots,80\ .$$

Concentrating on three time points we get the necessary set of simultaneous equations to solve for h_i, t_i, and k_i. This can be carried out for each segment i.

We can carry out this procedure even if the equation governing the dipole strength function (Eq. (3.3)), is much more complicated.

4. Computational Results

A number of simulated computer experiments for the determination of parameters h_i, k_i and t_i were performed, using the spline approximation as detailed in the text. This procedure is very simple and straightforward for cases where the number of observation points is the same as the number of dipole moments (i.e. $M = N$). A small on-line computer like the PDP-8 can be used to produce the results on a real-time basis. Judging from our accumulated examples, this method produces all the unknown parameters in the inverse problem with a very good precision.

In what follows, we provide two typical examples of computer experimentation for $M = N = 3$ and $M = N = 5$. As we mentioned before, we could actually consider all 20 myocardial segments and determine the dipole parameters with good accuracy. The positions x_i, y_i and z_i of these segments and ℓ, m and n which are the direction cosines of the dipole moment were taken from the simulation experiments. Segments numbered 2, 11 and 18 were considered for $N = 3$. The above three segments and segments 8 and 14 were included for experimentation in the case $N = 5$. The position coordinates a_j, b_j and c_j of the observation points for the case $N = 5$ were (15,0,0), (0,15,0), (0,0,15), (15,0,15), and (0,15,15) centimeters. The first three of these observation points were used in the case $N = 3$. The reference point was taken to be at infinity.

In order to produce a set of surface potential a Gaussian fit to the experimental dipoles was assumed. The values of parameters k_i, t_i

and h_i corresponding to these Gaussian curves are given in the first three columns of Table I. These values are considered as "experimental" values of the unknown parameters and the results of our computation must be compared to these values. The simulated skin potential at the corresponding observation point for the two cases of $M = 5$ and $M = 3$, are given in Figs. 9.2 and 9.3 respectively. Using these values of $b_j(t)$, the time histories of the dipole moments in a given segment, was computed from Eq. (3.1) and are given in Fig. 9.4 for both $M = 5$ and $M = 3$. Then, cardinal splines were constructed to represent these time histories and the unknown parameters h_i, k_i and t_i were computed, as detailed in the text. The computed values of the parameters for $M = 5$ are given in Table I. In the case of $M = 3$, the values for the corresponding dipoles, are very much the same as these given in Table I. This result exhibits excellent agreement with "experimental parameter values" (the first three columns of Table I) and shows that the inverse problem can be solved without dependence on the initial guess.

TABLE I. Assumed Parameters of the Dipole Moment Histories and Those Computed in the Inverse Problem

Dipole number	"Experimental" values k_i	t_i	h_i	Computed values k_i	t_i	h_i
2	1.15E-2	29.0	4.30E-2	1.15E-2	28.0	3.36E-2
8	6.37E-3	37.3	1.20E-1	6.37E-3	36.3	1.52E-1
11	5.42E-3	45.5	3.68E-2	5.42E-3	44.5	4.69E-2
14	7.07E-3	60.5	2.39E-5	7.07E-3	59.5	3.65E-5
18	2.04E-2	48.8	1.39E-10	2.04E-2	47.8	3.71E-10

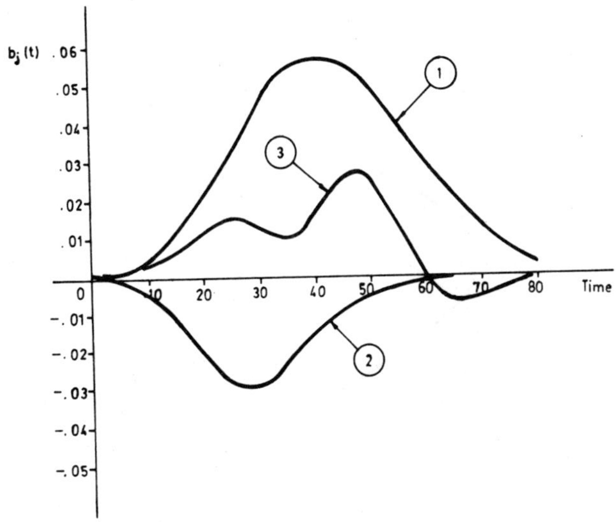

Fig. 9.3 Skin Potentials at the Observation Points for $M = 3$.

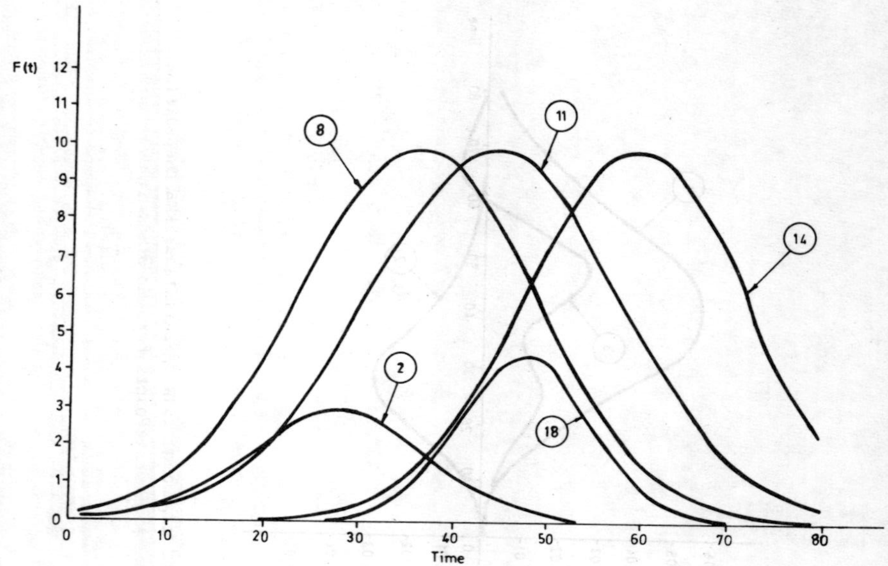

Fig. 9.4 Computed Dipole History for Both M = 3 and M = 5.

BIBLIOGRAPHY AND COMMENTS

1. Section 1. The following references will be useful:

   i. A.M. Scher and A.C. Young, "The Pathways of Ventricular Depolarization in Dogs", Circular Res., 4, 461 (1956).

   ii. D.D. Durrer, R.T. Van Dam, F.L. Meijler, R.C. Arzbacher, E.J. Mueller, and G.E. Freud, "Electrical Activation and Membrane Action Potential of a Perfused Normal Human Heart", Read before the American Heart Association in New York, Oct. 1966, Circulation, 34, 92 (1966).

   iii. R.H. Sylvester, R. Kalaba, C.R. Collier, R. Bellman, and H. Kagiwada, "A Digital Computer Model of the Vectorcardiogram with Distance and Boundary Effects: Simulated Myocardial Infarctions", American Heart Journal, 74, 792 (1967).

   iv. R.H. Sylvester, C.R. Collier and R.B. Pearson, "Analog Computer Model of the Vectorcardiograms", Circulation, 31, 45, (1965).

   v. R. Bellman, C. Collier, H. Kagiwada, R. Kalaba, and R. Sylvester, "Estimation of Heart Parameters Using Skin Potentials", Comm. Assoc. for Computing Machinery, 7, 666 (1964).

   vi. R. Bellman, B.G. Kashef and R. Vasudevan, "Differential Quadrature Partial Differential Equations and Splines", to appear.

   vii. R. Bellman, H. Kagiwada, and R. Kalaba, "Orbit Determination as a Multi-point Boundary Value Problem and Quasilinearization", Proc. Nat. Aca. Sci., 48, 1827 (1962).

   viii. F.N. Wilson and R.H. Bailey, "The Electric Field of an Eccentric Dipole in a Homogeneous Spherical Conducting Medium", Circulation, 1, 84 (1950).

   ix. R. Bellman, Introduction to Matrix Analysis, 2nd edition, (McGraw-Hill Book Company, New York and London, 1970).

   x. J.H. Ahlberg, E.N. Nelson and J.L. Walsh, The Theory of Splines and their Applications, (Academic Press, New York and London, 1967).

   xi. B.H. Kashef, R. Vasudevan, and R. Bellman, "Splines via Dynamic Programming", J. of Math. Anal. and Appl., 38, (1972).

2. Note that we are using the finite element method found in the book by N. Distefano, Nonlinear Processes in Engineering, (Academic Press, Inc., New York, 1974).

Chapter X

DYNAMIC PROGRAMMING AND AN INVERSE PROBLEM IN TRANSPORT THEORY

1. Introduction

The theory of neutron transport and the theory of radiative transfer are devoted to problems of determining properties of radiation fields, given the properties of the radiation sources and the medium. Inverse problems, which have received little study, are those in which we seek to determine properties of the medium, given those of the incident radiation and the radiation fields.

In this chapter we initiate the study of inverse problems in transport theory from the point of view of dynamic programming. Our goal is to produce a feasible computational method for estimating the properties of a medium based upon measurements of radiation fields within the medium. It is assumed that the reader is somewhat familiar with the invariant imbedding approach to transport theory.

For ease of exposition we consider a one-dimensional transport process. The method described here can be generalized to the vector-matrix case, and thus to the slab geometry with anisotropic scattering, to wave propagation, and to transmission lines.

2. Formulation

Consider the one-dimensional medium shown in Fig. 10.1.

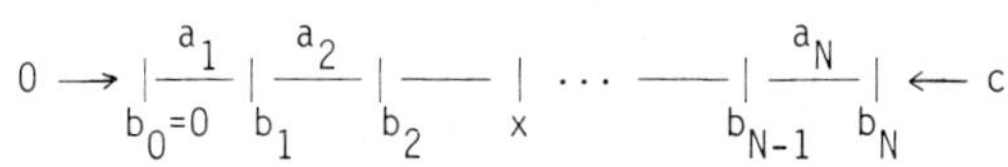

Fig. 10.1 A One-Dimensional Transport Process

It consists of N homogeneous sections (b_i, b_{i+1}), $i = 1,2,\ldots,N-1$. When a neutron travels through a distance Δ in the i-th section, there is a probability $a_i\Delta$ that it will interact with the medium. The result of an interaction is that the original neutron is absorbed and two daughter neutrons appear, each traveling in one direction. Suppose that c neutrons per unit time are incident from the right and zero neutrons per unit time are incident from the left. We denote the average number of particles per unit time passing the point x and moving to the right by $u(x)$ and the same quantity for the leftward moving particles by $v(x)$. Suppose that measurements on the interval intensities are made at various points $x = x_i \neq b_j$; for example,

$$(2.1) \qquad u(x_i) \cong w_i , \qquad i = 1,2,\ldots,M .$$

Our aim is to estimate the characteristics of the medium, the quantities a_i, $i = 1,2,\ldots,N$, on the basis of these observations.

The interval intensities satisfy the differential equations

$$(2.2) \qquad \frac{du}{dx} = a_i v , \qquad - v = a_i u$$

$$(2.3) \qquad b_{i-1} \le x \le b_i , \qquad i = 1,2,\ldots,N .$$

The analytical solution is of no import, since we wish to consider this as a prototype of more complex processes for which a computational treatment is mandatory. In addition, $u(x)$ and $v(x)$ are continuous at the interfaces; that is,

$$(2.4)\qquad u(b_i - 0) = u(b_i + 0)$$

$$(2.5)\qquad v(b_i - 0) = v(b_i + 0)\ , \qquad i = 1,2,\ldots,N-1\ ,$$

and

$$(2.6)\qquad u(0) = 0\ ,$$

$$(2.7)\qquad v(b_N) = c\ .$$

We wish to select the N constants $a_1, a_2, \ldots, a_N$, so as to minimize the sum of the squares of the deviations S,

$$(2.8)\qquad S = \sum_{i=1}^{M} \{u(x_i) - w_i\}^2\ .$$

3. Dynamic Programming

Let us suppose that the functions u and v are subject to the conditions of Sec. 2 and

$$(3.1)\qquad u(b_K) = c_1\ ,$$

$$(3.2)\qquad v(b_K) = c_2\ .$$

In addition we write

$$(3.3)\qquad f_K(c_1,c_2) = \min \sum_{i=1}^{M_K} \{u(x_i) - w_i\}^2\ ,$$

where the minimization is over the absorption coefficients $a_1,a_2,\ldots,a_K$. The number of observations on the first K interval is M_K. We view K as a parameter taking on the values $1,2,\ldots,$ and c_1 and c_2 are also viewed as variables. Then we write

$$(3.4)\qquad f_1(c_1,c_2) = \sum_{i=1}^{M_1} \{u(x_i) - w_i\}^2\ ,$$

where

(3.5) $$u(b_1) = c_1 \quad ,$$

(3.6) $$v(b_1) = c_2 \quad ,$$

(3.7) $$\frac{du}{dt} = a_1 v \quad ,$$

(3.8) $$\frac{dv}{dt} = a_1 u \quad .$$

The absorption coefficient a_1 is chosen so that

(3.9) $$u(0) = 0 \quad .$$

In addition, the principle of optimality yields the relationship

(3.10) $$f_{K+1}(c_1,c_2) = \min_{a_{K+1}} \{d_{K+1} + f_K(c_1',c_2')\} \quad , \qquad K = 1,2,\ldots,$$

(3.11) $$d_{K+1} = \sum_i \{u(x_i) - w_i\}^2 \quad ,$$

with i ranging over integer values for which

(3.12) $$b_K < x_i < b_{K+1} \quad ,$$

and

(3.13) $$\frac{du}{dx} = a_{K+1} v, \qquad u(b_{K+1}) = c_1 \quad ,$$

(3.14) $$-\frac{dv}{dx} = a_{K+1} u, \qquad v(b_{K+1}) = c_2 \quad .$$

In addition we have

(3.15) $$c_1' = u(b_K) \quad ,$$

(3.16) $$c_2' = v(b_K) \quad .$$

In the usual manner of dynamic programming, this leads to a computational scheme for producing the sequence of functions of two variables $f_1(c_1,c_2)$, $f_2(c_1,c_2),\ldots,$ and in principle solves our estimation problem. In the event that we do not wish to require that $u(0) = 0$, we may determine the function $f_1(c_1,c_2)$ in this way:

$$(3.17) \qquad f_1(c_1,c_2) = \min_{a_1} \left[\lambda u^2(0) = \sum_{i=1}^{M_1} \{u(x_i) - w_i\}^2\right],$$

where λ is a suitably large parameter.

4. An Approximate Theory

While the original physical problem is a two-dimensional problem, it may be well represented as a one-dimensional problem. Suppose that there are K segments of the medium and that the input is $v(b_K) = c$. The absorption coefficients $a_1,a_2,\ldots,a_K$ should be chosen to secure a minimum sum of squares of deviations from the measurements. Having picked the absorption coefficients, we may calculate the reflection coefficient $r(v(b_K))$ for this segmented medium. At the end b_K, the function u is essentially determined by v and $r(v)$, $u(b_K) = v(b_K)\, r(v(b_K))$. The single variable $v(b_K) = c$ may then suffice to specify the state at the right end of the K-th segment.

Let us define the function $g_N(c)$,

(4.1) $\qquad g_N(c)$ = the smallest sum of squares of deviations on the first N segments when the input is c,

and the function $R_N(c)$,

(4.2) $\qquad R_N(c)$ = the reflection coefficient that results when the optimal set of absorption coefficients is used on the first N segments, the input being $c = v(b_N)$.

The function $g_{N+1}(c)$ satisfies the inequality

(4.3) $$g_{N+1}(c) \le \min_a \{d_{N+1} + g_N(c')\} ,$$

where

(4.4) $$d_{N+1} = \sum[u(x_i) - w_i]^2 , \qquad b_N < x_i < b_{N+1} ,$$

and

(4.5) $$\frac{du}{dx} = av , \qquad v(b_{N+1}) = c ,$$

(4.6) $$-\frac{dv}{dx} = au , \qquad v(b_N)R_N(v(b_N)) = u(b_N) ,$$

and

$$c' = v(b_N) .$$

We do not have recurrence relations for the sequences of functions $g_N(c)$ and $R_N(c)$. We replace Eqs. (4.3), (4.4) and (4.5) by an approximate set, where instead of $g_N(c)$ we introduce the sequence $f_N(c)$, and instead of $R_N(c)$ we introduce $r_N(c)$. In our approximate theory we produce $f_{N+1}(c)$ from the recurrence formula

(4.7) $$f_{N+1}(c) = \min_a \{d_{N+1} + f_N(c')\} ,$$

where

(4.8) $$d_{N+1} = \sum_i[u(x_i) - w_i]^2 , \qquad b_N < x_i < b_{N+1} ,$$

and

$$c' = v(b_N) .$$

The following boundary value problem,

$$\frac{du}{dx} = av , \qquad v(b_{N+1}) = c ,$$

$$(4.9)\qquad -\frac{dv}{dx} = au \ , \qquad v(b_N)r_N(v(b_N)) = u(b_N)$$

must be satisfied. The quantity

$$(4.10)\qquad r_{N+1}(c) = r(b_{N+1})$$

is obtained as the solution of the initial value problem

$$(4.11)\qquad \frac{dr}{dx} = a(1 + r^2) \ , \qquad r(b_N) = r_N(c')$$

For $N = 1$ we define

$$(4.12)\qquad f_1(c) = \min_a \sum_i \alpha[u(x_i) - w_i]^2 \ ,$$

where the summation is over indices for which

$$(4.13)\qquad 0 < x_i < b_1 \ ,$$

and α is a weighting constant. Also we have

$$\frac{du}{dx} = av \ , \quad u(0) = 0 \ ,$$

$$(4.14)\qquad -\frac{dv}{dx} = au \ , \qquad v(b_1) = c \ .$$

We define

$$(4.15)\qquad r_1(c) = r(b_1)$$

where

$$r = a(1 + r^2) \ , \qquad 0 \le x \le b_1 \ ,$$

$$(4.16)\qquad r(0) = 0 \ .$$

The purpose of introducing the weight $\alpha \ge 1$ is to insure a good fit over the first segment.

Assuming that a unique minimizing solution exists, we can show that the results of our approximate theory are exact, if the observations w_i are perfectly accurate. We reason inductively. For the one-segment process, there exists an input c_1^* for which $f_1(c_1^*) = 0$ by Eq. (4.12), and the reflection coefficient is $r_1(c_1^*)$. We assume that there exists an input to the medium of N segments, c_N^*, such that $f_N(c_N^*) = 0$, and that the reflection coefficient for this medium is $r_N(c_N^*)$. For the medium of $N+1$ segments, there is an input c_{N+1}^* such that $d_{N+1} = 0$, and the input (to the left) at b_N which satisfies condition (4.10) is $v(b_N) = c_N^*$. Therefore $f_{N+1}(c_{N+1}^*) = 0$, and the solution is exact.

In this manner we have reduced the original multi-dimensional optimization process to a sequence of one-dimensional processes.

5. A Further Reduction

The solving of the nonlinear boundary value problem of Eqs. (4.5) and (4.9) can be a source of difficulty. To aid in this process we note that we can write

$$v(b_N) = cT + u(b_N)R \quad , \tag{5.1}$$

which follows simply from one of Chandrasekhar's invariance principles. The transmission coefficient T and the reflection coefficient R of the $(N+1)$st segment are calculated from the solutions of the initial value problems

$$\frac{dr}{dx} = a(1 + r^2) \ , \qquad r(0) = 0 \tag{5.2}$$

$$\frac{dt}{dx} = art \ , \qquad t(0) = 1 \quad , \tag{5.3}$$

and

$$R = r(b_{N+1} - b_N) \quad , \tag{5.4}$$

$$T = t(b_{N+1} - b_N) \quad . \tag{5.5}$$

In this way the second boundary condition in (4.9) becomes

$$r_N(v)v = (v - cT)/R \quad , \tag{5.6}$$

a nonlinear equation for $v = v(b_N)$. The other condition in (4.9) is retained.

6. Computational Procedure

The calculation of f_{N+1} for a given value of the parameter c may proceed as follows. We take a value of the coefficient a, and we produce numerically the reflection and transmission coefficients R and T. Assuming we can solve Eq. (5.6) for $v(b_N)$, we go on to solve the linear two-point boundary value problem of Sec. 3 by producing numerically two independent solutions of these homogeneous equations and determining constants so that the boundary conditions are met. Then the sum of squares of deviations d is computed, and the cost $\{d + f_N(v(b_N))\}$ is evaluated. We go through these steps for all the admissible choices of a, and the costs are compared. The value of a which makes the cost a minimum, is the choice for the $(N+1)$st slab. The whole procedure is repeated for the range of values of c and of N.

It may be noted that in the calculation of the reflection coefficient r_{N+1}, the initial condition r_N is known only computationally on a grid of values of the argument. Experiments are needed to determine the required fineness of grid to achieve the required accuracy.

It is possible to derive recurrence relations for $f'_N(c)$ and $r'_N(c)$, and these can be employed in a variety of ways to improve the accuracy of the method. Numerical experimentation would have to be carried out to obtain reliable estimates of running times and accuracies. The method proposed here can be extended to treat the case where the interface points are not known, though the computational effort will be greatly increased.

Experience with many similar problems leads us to believe that the proposed procedure is perfectly feasible.

7. Computational Results

A. Production of Observations

We consider a homogeneous rod of unit length with absorption coefficient $a = 0.5$. We produce the internal fluxes to the right and to the left due to a unit input flux to the left at the end $x = 1$ and no input at the other end, $x = 0$. To do this, we use the fact that the quantity $u(1)$ is the reflection coefficient for the slab, which is $\tan a$. We integrate the transport equations with the initial values $v(1) = 1$, $u(1) = \tan a$, from $x = 1$ to $x = 0$. This procedure yields $u(x)$ and $v(x)$ throughout the rod.

B. Two-dimensional Dynamic Programming Procedure for the Determination of the Absorption Coefficients

The rod is divided into 10 homogeneous sections of equal length. From the set of exact measurements, $w_i \approx u(x_i)$, we wish to determine the set of optimizing parameters a_N in each section. The correct solution is $a_N = 0.5$ for $N = 1,2,\ldots,10$.

In stage one of the multi-stage decision process, the rod is considered to consist of one segment extending from $x = 0$ to $x = 0.1$. If $c_1 = u(0.1)$, $c_2 = v(0.1)$, we choose the coefficient which makes $u(0) = 0$,

$$a = \frac{1}{0.1} \text{ arc tan } \frac{c_1}{c_2} ,$$

regardless of the measurements in this segment. The minimum cost is $f_1(c_1,c_2) = \sum_i [u(x_i) - w_i]^2$, where $u(x_i) = \sin a\, x_i$, $0 < x_i < 0.1$. This calculation is carried out for each value of c_1 and c_2.

The computations for the other stages $N = 2,3,4,\ldots,$ may be best indicated by the following outline:

Two-dimensional Dynamic Programming Calculations

For each stage $N = 2,3,4,\ldots,$

1. Print N
2. For each c_1
   1. For each c_2
      1. For each a
         1. Integrate to produce $c_1' = u(b_{N-1})$, $c_2' = v(b_{N-1})$,
            $$\begin{cases} \dot{u} = a\,v\,, & u(b_N) = c_1 \\ -\dot{v} = a\,u\,, & v(b_N) = c_2 \end{cases}$$
         2. Compute $d = \sum_i [u(x_i) - w_i]^2$
         3. Find $f_{N-1}(c_1',c_2')$ by interpolation
         4. Set $S(a) = d + f_{N-1}(c_1',c_2')$
      2. Search for $f_N(c,c) = \min_a \{S(a)\}$
      3. Print $c_1, c_2, a_N, c_1', c_2', f_N(c_1,c_2)$
3. For each c_1
   1. For each c_2
      1. Shift $f_N(c_1,c_2) \to f_{N-1}(c_1,c_2)$

There are four levels of computation: the stage N, the state c_1, the state c_2, and the parameter a. The large brackets cover the steps which must be carried out at each level. The statement "shift $f_N \to f_{N-1}$" represents the discarding of the costs for stage N-1, and the replacement of f_{N-1} by the just computed costs for stage N, to get ready for the next stage. This saving in storage is allowed by the recurrence formula linking the current cost with the cost of only the previous stage. The interpolation may be carried out by the use of a linear formula in two dimensions, c_1 and c_2. The printout value of a is, of course, the optimal value.

In our numerical trial, we execute the algorithm for three stages only, the rod then extending from $x = 0$ to $x = 0.3$. The exact observations are:

$$u(0.02) = 0.11394757 \times 10^{-1}$$
$$u(0.05) = 0.28484388 \times 10^{-1}$$
$$u(0.08) = 0.45567610 \times 10^{-1}$$
$$u(0.12) = 0.68328626 \times 10^{-1}$$
$$u(0.15) = 0.85381951 \times 10^{-1}$$
$$u(0.18) = 0.10241607 \times 10^{0}$$
$$u(0.22) = 0.12509171 \times 10^{0}$$
$$u(0.25) = 0.14206610 \times 10^{0}$$
$$u(0.28) = 0.15900853 \times 10^{0}$$

The range of N is 1 to 3, the section interfaces lying at $x = 0.1$, 0.2, 0.3. The range of c_1 is 0.00 (0.01) 0.20, 21 values; the range of c_2 is 1.120 (0.002) 1.140, 11 values. The five allowed values of a are 0.3 (0.1) 0.7. From the direction calculation, that is, when the true structure of the rod is given, we know the conditions at the right end ($x = 0.3$) are $u(0.3) = 0.17028385$, $v(0.3) = 1.1266986$. The inverse calculations do not produce clearly the correct results $a_1 = a_2 = a_3 = 0.5$. It is believed that the grids of values of c_1 and of c_3 are not sufficiently fine, and that substantially improved results cannot be obtained without a great increase in computing expense. The one-dimensional reduction appears attractive in view of these results.

C. One-dimensional Dynamic Programming Approximation for the Determination of the Absorption Coefficients

The rod of unit length is divided into five sections of equal length 0.2. Armed with the internal measurements $w_i \approx u(x_i)$, we wish to determine the absorption coefficient of each slab. The correct choices are $a_N = 0.5$ for $N = 1,2,\ldots,5$. In the one-dimensional case, the only state variable is $c = v(b_N)$.

The following outline lists the calculations for producing a_1, $f_1(c)$ and $r_1(c)$ for $N = 1$, and the next outline shows the general scheme, $N = 2,3,\ldots,$.

<u>One-dimensional Dynamic Programming</u>

For stage $N = 1$

1. Print N
2. For each $c = v(b_N)$
   1. For each a
      1. Solve 2 point boundary value problem for $v(0) = c$,
         $$\begin{cases} \dot{u} = a\,v\,, & u(0) = 0 \\ -\dot{v} = a\,u\,, & v(0.2) = c \end{cases}$$
      2. Integrate to produce $u(x)$
         $$\begin{cases} \dot{u} = a\,v\,, & u(0) = 0 \\ -\dot{v} = a\,u\,, & v(0) = c' \end{cases}$$
         and simultaneously calculate $d = \sum_i [u(x_i) - w_i]^2$,
         and keep a running estimate of $f_1(c) \approx \min_a \{d\}$.
   2. Integrate to produce $r_1(c) = \rho(0.2)$
      $\dot{\rho} = a(1 + \rho^2)$, $\rho(0) = 0$
   3. Print c, a_1, c', $r_1(c)$, $f_1(c)$

One-dimensional Dynamic Programming

For each stage $N = 2,3,4,\ldots$

1. Print N
2. For each c
    1. For each a
        1. Produce $R = \rho(0.2)$, $T = t(0.2)$ by integration
        $$\begin{cases} \dot{\rho} = a(1 + q^2), & \rho(0) = 0 \quad , \\ \dot{t} = a \rho t, & t(0) = 1 \quad . \end{cases}$$
        2. Solve the nonlinear equation for $c' = v(b_{N-1})$ and $r_{N-1}(c')$,
        $$Rc'r_{N-1}(c') = c' - cT \quad .$$
        3. Solve 2 point boundary value problem for $e' = u(0) = u(b_{N-1})$
        $$\dot{u} = a v , \quad v(0) = c'$$
        $$-\dot{v} = a u , \quad v(0.2) = c$$
        4. Integrate to produce $u(x)$
        $$\begin{cases} \dot{u} = a v , & u(0) = e' \\ -\dot{v} = a u , & v(0) = c' \end{cases}$$
        and calculate $d = \sum_i [u(x_i) = w_i]^2$.
        5. Find $f_{N-1}(c')$ by interpolation.
        6. Set $S(a) = d + f_{N-1}(c')$
        and keep a running estimate of $f_N(c) \simeq \min_a \{S(a)\}$.
    2. Integrate to produce $r_N(c) = \rho(0.2)$,
    $$\dot{\rho} = a(1 + \rho^2), \quad \rho(0) = r_{N-1}(c') \quad .$$
    3. Print c, a_N, c', $r_N(c)$, $f_N(c)$.
3. For each c
    1. Shift $f_N(c) \rightarrow f_{N-1}(c)$.

To solve the nonlinear equation (4.7) for $c' = v(b_{N-1})$, where $r(c')$ is known only on a grid of points, we compute the expressions $g_1 = Rc'r_{N-1}(c')$, $g_2 = c' - cT$, and we take their difference, $D = g_1 - g_2$. If $D = 0$, then c' has been found. Otherwise we repeat the procedure for each discrete value c_1', until the sign of D_i is opposite to that of D_{i-1}. We then interpolate linearly to find the quantity c' which makes $D = 0$. If the sign of D does not change, that is, the curves g_1 and g_2 do not intersect, then the corresponding value of a is definitely not allowed to be the coefficient for the segment in question.

If the minimum cost $f_N(c)$ is large for a given state c, this state and all remaining states may be delected from further consideration. This provides a saving in computing time, for each state to be considered requires many calculations. Of course, the precaution must be taken to order the c's properly, so that no potentially vital state is lost.

The proposed one-dimensional scheme has been tested numerically. The range of N is 1 to 3, the interfaces of the sections being located at $x = 0.2$, 0.4 and 0.6. The states $v(b_N) = c$ are 1.04955 (0.00015) 1.13385, 563 in all. This number is reduced in stage two to 546, by the use of the above test. Four values of the absorption coefficient a are allowed: 0.1, 0.3, 0.5, and 0.7. There are nine perfectly accurate observations of u per segment, a total of 27 data points. The integration method is Adams-Moulton with a grid size of 0.01.

From the output of our computation, we see that the minimum value of $f_3(c)$ is 0.387×10^{-8} and occurs when the input flux is $v(0.6) = c = 1.08855$ and the absorption coefficient for the segment of the medium between $x = 0.4$ and $x = 0.6$ is taken to be $a = 0.5$. This is very close to the true answer, $v(0.6) = 1.08860$, and the value of the parameter a is correct. The calculation tells us that the next state at $x = 0.4$ will be $v = 1.11673$. The nearest grid point in c is 1.11675, and the cost $f_2(1.11675)$ is indeed a minimum, 0.824×10^{-9}. The absorption coefficient for segment 2 is 0.5, the correct solution. The next state at $x = 0.2$ is predicted to be 1.13377. The nearest discrete state is 1.13385, possessing a cost $f_1(1.13385) = 0.181 \times 10^{-9}$. The absorption coefficient is 0.5 which is again the correct answer. The solutions at each stage are clearly found, the minimum cost differing from the others by at least an order of magnitude. These dynamic programming calculations have very accurately determined the input, and they have identified the medium.

Now we wish to test the one-dimensional method of determining the structure of the medium when the measurements are few and of limited accuracy. We consider the rod of length 0.8 consisting of four segments of equal length 0.2. There are again the same 563 discrete states in c, and the same four possible absorption coefficients 0.1, 0.3, 0.5 and 0.7. However, there are only three observations per segment and these are correct to only two significant figures. Knowing the inputs to the first three stages $N = 1,2,3$, we see from the output of the calculations that the absorption coefficients are $a_1 = a_2 = a_3 = 0.5$, the correct solution in this region. On the other hand, we are not able to accurately identify the input to a given segment on the basis of these calculations, because the minimum of the function f is broad and it is not centered at the correct value of the input c. For stage $N = 4$, the value of a_4 is determined to be 0.3, an incorrect value. These experiments might serve as a warning to the experimental investigator. They show that the processing of data with a small number of measurements requires higher accuracy than two figures, and that if the measurements are of limited accuracy, many measurements should be made.

BIBLIOGRAPHY AND COMMENTS

1. Section 1. We are following the paper by R. Bellman, H. Kagiwada and R. Kalaba, "Dynamic Programming and an Inverse Problem in Neutron Transport Theory", Computing, 2, Fasc. 1, 5-16 (1967).

2. For the theory of invariant imbedding, see the book, R. Bellman and G.M. Wing, An Introduction to Invariant Imbedding, (John Wiley & Sons, Inc., New York, 1974).

Chapter XI

AN INVERSE PROBLEM IN MATHEMATICAL PHYSICS

1. Introduction

The problem of solving the linear partial differential equation of diffusion theory and heat conduction,

$$\phi(x)\,\frac{\partial u}{\partial t} = \frac{\partial^2 u}{\partial x^2}\,, \tag{1.1}$$

subject to appropriate initial and boundary conditions has been extensively investigated using a variety of analytic techniques. We wish to consider a type of problem associated with this equation which has not been investigated to any extent vis-a-vis the inverse problem.

Instead of assuming that the coefficient function $\phi(x)$ is known and trying to find the solution $u(x,t)$, we shall suppose that the solution is known at certain points of the x,t plane, and try to determine the function $\phi(x)$ from this information. This type of question appears first to have been asked by the mathematical physicist, Ambarzumian, and it was also posed by Langer, who gave an analytic procedure for determining the coefficient function. The first detailed analysis of this problem was given by Borg, where references to earlier work by Langer and others may be found. Similar questions arise in the quantum theory of scattering, and were, in principle, solved by Gelfand and Levitan. In the paper of Kay, references to work of Levinson and others may be found.

The aim of this chapter is to show how problems of this nature are closely related to some classical synthesis problems for electrical networks. Our new approach furnishes a way of obtaining approximate solutions which appear to be more useful from the computational, engineering and physical point of view than the exact solutions furnished implicitly by the methods of Borg and Gelfand and Levitan. Here, we only sketch the method.

2. The Analytic Problem

Consider Eq. (1.1) holds for $0 < x < L$, $t > 0$, subject to the initial and boundary conditions,

$$u(L,t) = 0 \quad , \tag{2.1a}$$

$$u(x,0) = 0 \quad , \tag{2.1b}$$

$$\left.\frac{\partial u}{\partial x}\right|_{x=0} = \left[Z\left(\frac{d}{dt}\right)u(x,t)\right]_{x=0} \quad . \tag{2.1c}$$

Here $Z(s)$ is a meromorphic function of s which we shall specify in further detail below. Taking Laplace transforms with respect to t, we obtain the Sturm-Liouville problem

$$s\phi(x)v = \frac{d^2v}{dx^2} \quad , \tag{2.2a}$$

$$v(L,s) = 0 \quad , \tag{2.2b}$$

$$\left.\frac{dv}{dx}\right|_{x=0} = Z(s)v\Big|_{x=0} \quad , \tag{2.2c}$$

where

(2.3) $$v(x,s) = \int_0^\infty u(x,t)e^{-st}\,dt \ .$$

Given the function $Z(s)$, we wish to determine the density function. For an appropriate choice of $Z(s)$, this problem is equivalent to the problem considered by Borg. To see this, suppose that $Z(s)$ has the form

(2.4) $$Z(s) = \frac{\prod_{n=1}^{\infty}(1+s/\lambda_n)}{\prod_{n=1}^{\infty}(1+s/\mu_n)} \ ,$$

where λ_n, $\mu_n > 0$. If $s = -\lambda_n$, we have the Sturm-Liouville problem

$$v'' + \lambda_n v = 0 \ ,$$

(2.5) $$v'(0) = v(L) = 0 \ .$$

If $s = -\mu_n$, we have the problem

$$v'' + \mu_n v = 0 \ ,$$

(2.6) $$v(0) = v(L) = 0 \ .$$

We thus have the two distinct problems required for the solution of our problem following Borg's methods. Unfortunately, Borg's results do not furnish any useful technique for actually obtaining the unknown function, $\phi(x)$. We shall present another approach which thus furnishes an alternative attack upon Borg's problem.

3. Discrete Approximation

Let us, following a standard route, replace the differential equation of (2.2a) by the difference equation

(3.1) $$v_{K+1} - 2v_k + v_{k-1} = \Delta^2 s\phi_k v_k \ , \qquad k = 1,2,\ldots,N-1 \ ,$$

where

$$v_k = v(k\Delta) \ , \qquad \phi_k = \phi(k\Delta) \ , \qquad N\Delta = L \tag{3.2}$$

The boundary conditions of Eqs. (2.2b) and (2.2c) are replaced by

$$v_N = 0 \ , \tag{3.3a}$$

$$v_1 - v_0 = \Delta Z_N(s) v_0 \tag{3.3b}$$

We replace the meromorphic function $Z(s)$ by its approximation

$$Z_N(s) = \frac{\prod_{n=1}^{N} (1+s/\lambda_n)}{\prod_{n=1}^{N} (1+s/\mu_n)} \tag{3.4}$$

4. Network Theory

Once the problem has been transformed into that of determining the values of ϕ_k, an approximation to the determination of $\phi(x)$, we see that it is equivalent to that of determining the elements of a two-port network, a network of transmission-line type, given the response function.

Probably the most efficient and convenient way of deriving the values of the ϕ_k is based upon the use of continued fraction expansions. The details of this procedure, together with various necessary and sufficient conditions upon $Z(s)$ and the zeros and poles, $-\lambda_n$ and $-\mu_n$, may be found in Guillemin, Weinberg, Weinberg and Slepian, where many further references may be found.

Consider the system of equations

$$v_{k+1} - (2-\Delta^2 s\phi_k)v_k + v_{k-1} = 0 \ , \qquad k = 1,2,\ldots,N-1 \ . \tag{4.1}$$

Writing

$$(4.2)\qquad \frac{v_{k-1}}{v_k} = (2-\Delta^2 s\phi_k) - \frac{1}{(v_k/v_{k+1})} ,$$

we see that we obtain the finite continued fraction

$$(4.3)\qquad \frac{v_0}{v_1} = (2-\Delta^2 s\phi_1) - \frac{1}{(2-\Delta^2 s\phi_2)} .$$

Hence, any method which permits us to write v_0/v_1, as given by Eq. (3.3b), as a continued fraction of this form, yields the values ϕ_1, $\phi_2,\ldots,\phi_N$, and thus gives us a computational solution of the discrete inverse problem.

That there is a strong connection between the mathematical problems of network theory and those of quantum mechanics is no longer surprising in view of the work of Wigner and others on positive real functions.

BIBLIOGRAPHY AND COMMENTS

1. Section 1. We are following the paper by R. Bellman and J.M. Richardson, "A Note on an Inverse Problem in Mathematical Physics", Quarterly of Applied Mathematics, 19, 269-271 (1961).

2. For the reading of this chapter the following references will be useful:

   i. G. Borg, "Eine Umkehrung der Sturm-Liouvilleschen Eigenwerte Aufgabe", Acta Math., 78, 1-96 (1946).

   ii. I.M. Gelfand and B.M. Levitan, "On the Determination of a Differential Equation from its Spectral Function", Trans. Amer. Math. Soc., 2, 253-304 (1955).

   iii. I. Kay, "The Inverse Scattering Problem when the Reflection Coefficient is a Rational Function", Comm. Pure Appl. Math., 13, 371-406 (1960).

   iv. E.A. Guillemin, "A Summary of Modern Methods of Network Synthesis", Advances in Electronics, Academic Press Inc., 3, 261-303 (1951).

   v. L. Weinberg, Linear Network Synthesis, (McGraw-Hill Book Co., Inc., New York), to appear.

vi. L. Weinberg and P. Slepian, "Realizability Conditions on N-port Networks", IRE Trans. on Circuit Theory, CT-5, 217-221 (1958).

vii. E. Wigner, "On a Class of Analytic Functions from the Quantum Theory of Collisions", Ann. Math., 53, 36-67 (1951).

viii. A.M. Lane and R.G. Thomas, "R-matrix Theory of Nuclear Reactions", Rev. Mod. Phys., 30, 257-352 (1958).

Chapter XII

THE IDENTIFICATION OF SYSTEMS WITH PARTIAL INFORMATION

1. Introduction

In this chapter we consider the problems of identifying a system when only partial information is available to the observer.

In the problems discussed so far, we have assumed that the necessary observed data was available for use in the identifying techniques. In many instances this is not so and one must supply additional techniques to cope with this problem.

We also intend to discuss the interesting problems associated with identifying systems which are governed by partial differential equations. In this situation we will be interested in identifying not only the system parameters but also the initial conditions which now spread over a finite boundary of the system under investigation.

The systems which will be considered will be those governed by the partial differential equation,

$$\frac{\partial u(x,t)}{\partial t} = f(x,t,u,(x,t), \ \frac{\partial}{\partial x} u(x,t))$$

$$(1.1) \qquad a \leq x \leq b \ , \qquad t > 0.$$

We shall assume that the initial conditions for Eq. (1.1) are, at least, partially known and the solution can be sampled experimentally at certain predetermined points within the semi-infinite strip, $a \le x \le b$, $t > 0$. The problem we must solve is to determine the unknown initial conditions (and therefore the unknown solutions) at certain points within the range of x if we know the governing differential equations and can sample the solution experimentally at a finite number of points.

When only partial information is available, then we sample over a fewer number of points x_i in the space domain, but over longer times. Clearly from an experimental standpoint such a sampling procedure would have practical applications. It is the purpose of this chapter to develop techniques for doing this and to test the validity of the method through some numerical experimentation.

2. An Iterative Method of Solution

In the following sections we will seek to determine a mathematical fit to the partial information which is available to us. In doing so we shall construct an iterative procedure which, beginning with an initial approximation, will converge to a solution best fitting the known data in the least squares sense.

The method which we shall use first approximates the governing differential equation of the system (1.1) by a set of ordinary differential equations in time, using the ideas of differential quadrature. These (possibly nonlinear) equations are further linearized by the technique of quasilinearization. At each iteration, quasilinearization permits us to compute a class of possible solutions to the particular problem under study. Finally the "observed" data which is known by sampling the solution at predetermined spatial points, is used to construct the best solution in the least squares sense.

The iterative procedure ends when suitable convergence is obtained.

3. Differential Quadrature

Differential quadrature is an approximation technique for reducing a partial differential equation to a set of ordinary differential equations. Consider the nonlinear first order differential Eq. (3.1).

$$\frac{du}{dt}(x,t) = g\left(x,t,u(x,t), \frac{du}{dx}(x,t)\right), \quad a \le x \le b, \quad t > 0 \tag{3.1}$$

with the initial conditions,

$$u(x,0) = h(x) \quad . \tag{3.2}$$

If we make the assumption that the function $u(x,t)$, satisfying Eqs. (3.1) and (3.2), is sufficiently smooth, then we can write,

$$\frac{du}{dx}(x_i,t) = \sum_{j=1}^{N} a_{ij} u(x_j,t) \ , \qquad i = 1,2,\ldots,N \tag{3.3}$$

where the coefficients a_{ij} must be determined by a suitable method.

Substituting Eq. (3.3) into Eqs. (3.1) and (3.2) yields a system of ordinary differential equations of the form

$$\frac{du}{dt}(x_i,t) = g\left(t,x_i,u(x_i,t) \ , \ \sum_{j=1}^{N} a_{ij} u(x_j,t)\right) \quad , \tag{3.4}$$

with the initial conditions,

$$u(x_i,0) = h(x_i) \ , \qquad i = 1,2,\ldots,N \quad . \tag{3.5}$$

The numerical solution of the type of system illustrated by Eqs. (3.4) and (3.5) is a simple task for modern computers and also because the system is an initial value problem.

4. Determination of the Weighting Coefficients

If we now consider the differential quadrature,

$$(4.1) \qquad u_x(x_i) = \sum_{j=1}^{N} a_{ij}\, u(x_j) \ , \qquad i = 1,2,\dots,N \ ,$$

we can demand that Eq. (4.1) be exact for all polynomials of degree less than or equal to N-1.

By letting the test functions be $p_k(x) = x^{k-1}$, $k = 1,2,\dots,N$ and selecting arbitrary but distinct x_i's, the system leads to a set of algebraic equations,

$$(4.2) \qquad \sum_{j=1}^{N} a_{ij}\, x_j^{k-1} = (k-1)x_i^{k-2} \ , \qquad k = 1,2,\dots,N \ ,$$

which has a unique solution since the coefficient matrix is a Vandermonde matrix.

Another more attractive approach is to consider the shifted Legendre polynomials of degree N, $P_N^*(x)$ (with the orthogonality range of (0,1)). These polynomials are defined in terms of the usual Legendre polynomials by the relation,

$$(4.3) \qquad P_N^*(x) = P_N(1-2x) \ .$$

We can use the Lagrange interpolation formula, as a convenient test function

$$(4.4) \qquad p_k(x) = \frac{L_n^*(x)}{(x-x_k)L_n'^*(x_k)}$$

where $L_N^*(x)$ is the shifted N-th order Legendre polynomial with the leading coefficient unity and x_k the k-th root of $L_N^*(x)$. Therefore $p_k(x_i) = \delta_{ki}$.

Using the fact that Eq. (4.1) is exact for $u(x) = p_k(x)$, we see that

$$p_k'(x_i) = \sum_{j=1}^{N} a_{ij} p_k(x_j) = a_{ik}$$

$$(4.5) \qquad a_{ik} = \frac{L_N'^*(x_i)}{(x_i - x_k)L_N'^*(x_k)} \quad , \qquad i \neq k \ .$$

In the case where $i = k$, the use of L'Hospital's rule plus the fact that $L_N^*(x)$ satisfies the differential equation,

$$(4.6) \qquad x(1-x)L_N^{*''}(x) + (1-2x)L_N^{*'}(x) + N(N+1)L_N^*(x) = 0 \quad ,$$

which gives

$$(4.7) \qquad a_{kk} = \frac{1-2x_k}{2x_k(x_k-1)} \quad .$$

5. Quasilinearization

By using differential quadrature, we have succeeded in transforming the partial differential Eq. (1.1) into a set of approximate ordinary differential equations,

$$(5.1) \qquad u_t(x_i,t) = g(x_i,t,u(x_i,t) \ , \ \sum_{j=1}^{N} a_{ij} \ u(x_j,t)) \quad ,$$

with the initial conditions,

$$(5.2) \qquad u(x_i,0) = b_i \ , \qquad i = 1,2,\ldots,N \quad ,$$

x_i's being the zeroes of the shifted N-th order Legendre polynomial.

Since the function g is generally nonlinear and the initial conditions may not be all known, we shall use the quasilinearization technique to establish a numerical solution.

By setting $u_i(t) = u(x_i,t)$, the ordinary differential equation associated with the point x_i has the form,

$$(5.3) \qquad \frac{du_i}{dt} = g(x_i,t,u_i, \ \sum_{j=1}^{N} a_{ij}u_j)$$

For each i we now introduce a sequence of functions $u_i^{(k)}(t)$, $k = 1,2,...$ which are generated in the following manner: If the set of functions $u_i^{(k)}(t) = 1,2,...,N$ are known for some k, then

$$\text{(5.4)} \qquad \frac{du_i}{dt}(K+1) = g_i(u^{(k)}) + \sum_{j=1}^{N} \frac{\partial g_i(u^{(k)})}{\partial u_j} (u_j^{(k+1)} - u_j^{(k)}) \, , \qquad i = 1,2,...,N \, .$$

If the initial approximation $u_i(t)^{(0)}$ is known, then (5.4) defines a sequence of functions.

Since Eq. (5.4) represents a system of linear differential equations, the solution can be represented as a particular solution plus one homogeneous solution.

Let,

$$\text{(5.5)} \qquad u_i^{(k+1)}(t) = p_i(t)^{(k+1)} + \alpha_i h_i(t)^{(k+1)} \, , \qquad i = 1,2,...,N$$

The particular solution is defined by the equation,

$$\frac{dp_i}{dt}(k+1) = g_i(u^{(k)}) + \sum_{j=1}^{N} \frac{\partial g_i(u^{(k)})}{\partial u_i} (p_i^{(k+1)} - u_i^{(k)})$$

$$\text{(5.6)} \qquad p_i(0)^{(k+1)} = 0 \, .$$

while the homogeneous solution is defined through the differential equation,

$$\frac{dh_i}{dt}(k+1) = \sum_{j=1}^{N} \frac{\partial g_i(u^{(k)})}{\partial u_j} (h_j^{(k+1)})$$

$$\text{(5.7)} \qquad h_i(0)^{(k+1)} = 1 \, .$$

From the definition of the initial conditions in (5.6) and (5.7), α_i in Eq. (5.5) can be quickly identified with the initial conditions of the solution at the point x_i.

If the initial conditions at x_i were known and if the system were linear, this iteration technique would yield a solution on the first iteration. If the system were nonlinear and the sequence converged, then it will converge to the solution of (5.1).

If however, some of the initial conditions were not known, then additional information must be used to obtain the solution. In the next section we shall explore ways in which the additional information can be obtained from the observed data.

6. The Character of the Partial Information

By partial information we mean that over the space-time domain of the problem the solution can be sampled experimentally only at certain distinct spatial points. To illustrate how this partial information can be used to determine the unknown initial conditions, we shall consider a simple example. Suppose the system under study is governed by the partial differential equation,

$$\frac{\partial u}{\partial t} = \frac{du}{dx}\, g(x,t,u,u_x) \quad . \tag{6.1}$$

Suppose for the moment $-\infty < x < \infty$ so there are no boundary conditions except the vanishing of the solution at infinity. Furthermore let the initial conditions be,

$$u(x,0) = b(x) \quad . \tag{6.2}$$

We shall further suppose that the system is such that the initial conditions $b(x)$ cannot be observed and therefore must be considered as unknown. Instead of the initial conditions which are required to obtain a solution, we assume that the process $u(x,t)$ can be observed at M points x_i, $i = 1,\ldots,M$, and over a time interval $t_1 \le t \le t_2$.

The set of data functions $r_i(t)$, $i = 1,\ldots,M$, $t_1 \le t \le t_2$ constitute partial information about the system. While this partial information will not completely describe the behavior of the system, this information can be used to "fit" the mathematical model to the "observed data".

Let N be the number of terms in the differential quadrature approximation to the system (3.3). Let x_i, $i = 1,2,\ldots,N$ be the zeroes of the N-th shifted Legendre polynomial $L_N^*(x)$. Further let $r_k(t)$, $k = 1,2,\ldots,M$ and N be known observations of the system taken over a subset of the zeroes of $L_N^*(x)$.

We seek to determine the solution at the remaining $(N-M)$ zeroes of $L_N^*(x)$. We shall assume that enough information is known of the system so that the slope $\frac{\partial u}{\partial x}$ can be determined at the unknown data points.

We require that, at the unknown data points, the difference between the observed slopes and the theoretical slopes be a minimum in the least squares sense.

Therefore, we define the measurement of error as,

$$J = \frac{1}{2} \sum_k \int_{t_1}^{t_2} \left[\left(\frac{\partial u_k}{\partial x}\right) - \sum_{i=1}^{N} a_{ki}\, u_i \right]^2 dt \tag{6.3}$$

where k is taken over the data points where the solution is unknown.

In Eq. (6.3), since the index i ranges over all data points, we can write,

$$\sum_{i=1}^{N} a_{ki}\, u_i = \sum_n a_{kn} u_n + \sum_m a_{km}\, (p_m + \alpha_m h_m) \tag{6.4}$$

where n ranges over all the data points where the solution is known and m ranges over those data points where the solution is unknown.

Therefore we have,

$$(6.5) \qquad J(\alpha_1 \ldots \alpha_m) = \frac{1}{2} \sum_k \int_{t_1}^{t_2} (G_k(t) - \sum_m a_{km} h_m(t) \alpha_m)^2$$

where

$$G_k(t) = \sum_n a_{kn} u_n(t) + \sum_m a_{km} p_m \quad .$$

The expression (6.5) can be written as a standard quadratic form,

$$(6.6) \qquad \Phi(\alpha_1 \ldots \alpha_M) = \sum_{i=1}^{M} \sum_{j=1}^{M} A_{ij} \alpha_i \alpha_j + \sum_{j=1}^{M} B_j \alpha_j + C \quad .$$

The set α_i, $i = 1,2,\ldots,M$ which minimizes Φ will enable us to define the $(k+1)$ iteration. Therefore using Eq. (4.5),

$$(6.7) \qquad u_i(t)^{(k+1)} = p_i(t)^{(k+1)} + \alpha_i h_i(t)^{(k+1)} \quad , \qquad i = 1,2,\ldots,M \quad .$$

Using the new solution $u_i(t)^{(k+1)}$, new particular and homogeneous solutions can be computed from Eqs. (4.6) and (4.7).

This process is continued until proper convergence of the set $(\alpha_i,\ i = 1,2,\ldots,M)$ is obtained. At that point the unknown solutions $u_i(t)$ can be generated by Eq. (6.7).

7. Numerical Experiments

As an illustrative example of the methods we have been discussing, we shall consider the hyperbolic nonlinear problem,

$$(7.1) \qquad \frac{du_t}{dt}(x,t) = u(x,t) \frac{du}{dx}(x,t) \ , \qquad 0 \le x \le 1 \ , \qquad 0 \le t \le T \ ,$$

with the initial condition,

$$(7.2) \qquad u(x,0) = g(x)$$

The system (7.1) has the implicit solution,

$$u(x,t) = g(x + ut) \tag{7.3}$$

which serves as an excellent check example. The shock phenomenon inherent in the solution of (7.1) can be pushed into the future by a suitable choice of the function $g(x)$ and will not be considered here.

Using the differential quadrature technique, Eq. (7.1) may be written as,

$$\frac{du_i}{dt} = u_i \sum_{j=1}^{N} a_{ij} u_j \quad , \qquad i = 1,2,\ldots,N \tag{7.4}$$

where $u_i(t) = u(x_i,t)$.

Since the constants a_{ij} are known (see Sec. 4) the quasilinear equations associated with (7.4) can be written down immediately as

$$\frac{du_i^{(k+1)}}{dt} = \sum_{j=1}^{N} a_{ij}(u_i^{(k)}u_j^{(k+1)} + u_j^{(k)}u_i^{(k+1)} - u_i^{(k)}u_j^{(k)}) \; , \tag{7.5}$$

$$i = 1,2,\ldots,N \; .$$

From (7.5) the defining equations for the particular and homogeneous solution can be established.

For the first numerical problem, we let the initial conditions be $g(x) = \alpha x$. In this case the exact solution is

$$u(x,t) = \frac{x}{(\alpha^{-1} - t)} \quad . \tag{7.6}$$

The order of the differential quadrature scheme is selected to be 9. Therefore the data points are located at,

x_1 = 1.591941 E-2 x_6 = 6.621265 E-1

x_2 = 8.199715 E-2 x_7 = 8.066868 E-1

x_3 = 1.932987 E-1 x_8 = 9.180116 E-1

x_4 = 3.378780 E-1 x_9 = 9.840817 E-1

x_5 = 5.000000 E-1

The data will be known at every point except x_2 and x_4. In this example, the solution at x_2 and x_4 will be approximated from the available data. In Table I, the convergence of the initial values of both $u_2(t)$ and $u_4(t)$ are shown. From the initial conditions the solutions can be easily computed.

The "experimental" values of the slopes at x_2 and x_4 are simply given as,

$$\frac{du_i}{dx} = \frac{u_{i+1}(t) - u_{i-1}(t)}{x_{i+1} - x_{i-1}} \quad , \qquad i = 2,4. \tag{7.7}$$

The data was generated from (7.1) for $\alpha^* = 0.2$, while the initial approximation to $u_2(t)$ and $u_4(t)$ was computed using $\alpha_0 = 0.25$. The time integration was from $t = 0$ to $t = 0.745$.

The results of 7 iterations of the process are shown in Table I.

For the next case where we ask for three solutions occurring at the points x_2, x_4, and x_6, we select the order of the differential quadrature to be $N = 12$.

TABLE I

αx_2	αx_4
2.049928 E-2	8.44695 E-2
1.626684 E-2	6.271469 E-2
1.535499 E-2	6.379562 E-2
1.516099 E-2	6.403684 E-2
1.518572 E-2	6.408005 E-2
1.518783 E-2	6.407904 E-2
1.518770 E-2	6.407910 E-2
1.639943 E-2	6.75756 E-2

The data points now become,

x_1 = 9.290099 E-3 $\qquad$ x_7 = 5.626163 E-1

x_2 = 4.782482 E-2 $\qquad$ x_8 = 6.838779 E-1

x_3 = 1.151043 E-1 $\qquad$ x_9 = 7.938504 E-1

x_4 = 2.063308 E-1 $\qquad$ x_{10} = 8.844552 E-1

x_5 = 3.160852 E-1 $\qquad$ x_{11} = 9.527831 E-1

x_6 = 4.373831 E-1 $\qquad$ x_{12} = 9.903955 E-1 .

The results of 7 iterations are shown in Table II.

For the second numerical example, we let the initial condition be $g(x) = 0.2\ x^2$. In this case the exact solution is,

$$u(x,t) = \frac{(1 - (0.4)xt - (1 - (0.8)xt)^{\frac{1}{2}}}{(0.4)t^2} \tag{7.8}$$

Again selecting $N = 12$ and choosing x_4, x_6, x_8, and x_{10} to be the unknown data points, the results are given in Table III.

TABLE II

αx_2	αx_4	αx_6
1.1956 E-2	5.15827 E-2	1.09345 E-1
1.014126 E-2	3.832701 E-2	8.739626 E-2
1.025612 E-2	3.865226 E-2	8.779704 E-2
1.024132 E-2	3.868346 E-2	8.781075 E-2
1.026673 E-2	3.865596 E-2	8.782094 E-2
1.026667 E-2	3.865061 E-2	8.782410 E-2
1.027357 E-2	3.864514 E-2	8.782666 E-2
1.028081 E-2	3.866529 E-2	8.781991 E-2
9.5650 E-3	4.1226 E-2	8.7477 E-2

TABLE III

αx_4^2	αx_4^2	αx_3^2	αx_{10}^3
1.0643 E-2	4.7826 E-2	1.1693 E-1	1.9557 E-1
8.2227 E-3	2.8734 E-2	9.3876 E-2	1.2123 E-1
7.0029 E-3	3.2111 E-2	9.2007 E-2	1.2600 E-1
5.9205 E-3	3.4370 E-2	8.9436 E-2	1.3128 E-1
6.1325 E-3	3.3958 E-2	8.9938 E-2	1.3038 E-1
6.3021 E-3	3.3644 E-2	9.0355 E-2	1.2967 E-1
6.2705 E-3	3.3701 E-2	9.0275 E-2	1.2980 E-1
8.5145 E-3	3.8261 E-2	9.3538 E-2	1.5615 E-1 *

*Exact value

For the final numerical example we let the initial condition be $u(x) = 0.2 \sin \pi x$. In this case the analytical solution is

(7.9) $u(x,t) = 0.2 \sin \pi(x + ut)$.

Selecting $N = P_2$ and choosing x_3 and x_6 to be the unknown data points, we get the results shown in Table IV, after 4 iterations.

TABLE IV

$\alpha \sin \pi x_3$	$\alpha \sin \pi x_6$
6.1706 E-2	1.5000 E-1
8.3532 E-2	1.9590 E-1
8.4209 E-2	1.9687 E-1
8.4265 E-2	1.9684 E-1
8.2306 E-2	2.0000 E-1 *

*Exact value

By keeping the same $N = 12$, and choosing x_2, x_4 and x_6 on the unknown data points, the results are given in Table V.

TABLE V

$\alpha \sin \pi x_2$	$\alpha \sin \pi x_4$	$\alpha \sin \pi x_6$
2.6493 E-2	1.0287 E-1	1.5000 E-1
1.8178 E-2	1.3732 E-1	1.9519 E-1
1.9728 E-2	1.3549 E-1	1.9280 E-1
1.7782 E-2	1.3912 E-1	1.9088 E-1
3.5297 E-2	1.3712 E-1	2.0000 E-1 *

*Exact value

8. Discussion

An evaluation of the numerical results leads to the cautious conclusion that this method of determining "missing" values in a nonlinear system governed by partial differential equations is successful. Of particular interest is the accuracy by which the solutions are found. The sources of error in the method arise from two principal approximations. The first is the differential quadrature technique coupled with the quasilinear iterative process. The second is the number of "missing" values one wishes to find. As the order of the differential quadrature is increased, the more accurately the solution is approximated. On the other hand, as the number of "missing" values increases, the accuracy of the data fit is expected to diminish.

This interplay of different sources of error is illustrated by comparing the results in Table I with those of Table II. The first example uses a differential quadrature order $N = 9$ and calls for two solutions while the second example asks for three solutions using a more accurate quadrature order of $N = 12$.

Table III uses a more complicated set of initial conditions and shows the convergence properties when four solutions are asked for out of a set of $12 (N = 12)$. Tables IV and V compare the convergence in which N remains at 12, but at first two values and then three values are sought.

The non-uniform distribution of the error in the data fit which is particularly evident in Table III and Table V may be due to the simple computation of the slope, e.g. (7.7).

The identification of nonlinear systems is a difficult task. Each problem must be considered on its own merits. Within this context the results represent, we feel, a small step in the vast field of systems identification.

9. Control Theory

It is clear that a large number of analytic problems are created by systems with imperfect information. We want to point out that there are many interesting control problems also. For example,

1. How does one make decisions on the basis of imperfect information?
2. What further testing does one do to get more information?

References for considering these questions are given at the end of the chapter.

10. A Scanning Procedure

In this section we wish to study the problem of system identification when the system is governed by a partial differential equation and the observations are obtained by a scan in the space-time domain.

We begin by assuming that the system under study is governed by a partial differential equation of the form,

$$u_t(x,t) = f(x,t,u(x,t), u_x(x,t)) \quad ,$$

$$a < x < b \ , \qquad t > 0 \ . \tag{10.1}$$

We sample the output of the system over a path in the (x,t) space, that is by scanning,

$$d(s) = u(x(s), t(s)) \ , \qquad 0 < s < s_0 \ . \tag{10.2}$$

A scanning technique can be used when the system is observed along a zig-zag path in the domain $(a \leq x \leq b, \quad t > 0)$, see Fig. 12.1.

If the scan is digital and covers the domain accurately, then the information can be regrouped to give a set of data functions,

$$d_i(t) = u(x_i,t) \ , \qquad i = 1,2,\ldots,N \ , \tag{10.3}$$

where the set x_1 is a preselected group of points in the range $a < x_1 < x_2 <\ldots< x_N < b$.

One of the characteristics of this regrouping of the scan data is that the initial conditions of the system are not known experimentally (except at least at one point).

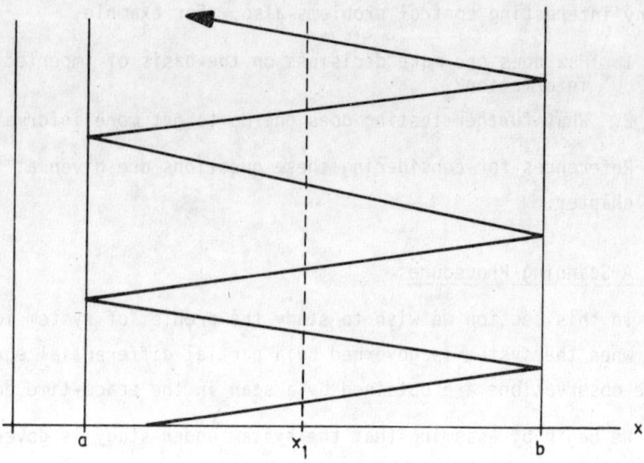

Fig. 12.1 A Zig-Zag Scanning Procedure

We now require that the scan data best fit the theoretical solution in the least squares sense.

11. Quasilinearization and the Estimation of Differential Operators from Eigenvalues

Iteration techniques for the determination of eigenvalues of ordinary differential equations are well developed. At least as interesting though, is the inverse problem: Given some properties of the eigenvalues, estimate some properties of the differential operator. In this section we show how the quasilinearization technique can be applied to this problem. The method is very promising for use in estimating the parameters of the potential function in a bound-state wave equation; for example, given the eigenfrequencies, determine the "force constants" in an oscillator problem.

Consider the differential equation

$$\frac{d^2u}{dx^2} + k^2(x)u = \lambda u , \qquad 0 \le x \le 1 , \tag{11.1}$$

together with the boundary conditions

$$(11.2)\qquad u(0) = u(1) = 0 \quad .$$

Let the first N eigenvalues be $\lambda_1, \lambda_2, \ldots, \lambda_N$, and let the first N eigenfunctions be $u_1(x)$, $u_x(x), \ldots, u_N(x)$. The eigenvalues may be determined.

Let $k^2(x)$ be viewed as a function of some parameters $a,b,c,\ldots$, as well as a function of x,

$$(11.3)\qquad k^2(x) = k^2(a,b,c,\ldots;\, x) \quad .$$

Suppose that we have obtained noisy estimates of the first N eigenvalues $\lambda_1, \lambda_2, \ldots, \lambda_N$. The best estimate of the differential operator reduces to the estimation of the function $k^2(a,b,c,\ldots;\, x)$. The problem is stated below in the form in which it will be considered.

Given the eigenvalues $\lambda_1, \lambda_2, \ldots, \lambda_N$, estimate the constants $a,b,c,\ldots$ such that S is minimized, $\min S$, where

$$(11.4)\qquad S = \sum_{n=1}^{N} u_n^2(1)$$

and

$$(11.5)\qquad u_n'' + k^2(a,b,c,\ldots;\, x)u_n = \lambda_n u_n \quad ,$$

$$(11.6)\qquad u_n(0) = 0 \quad ,$$

$$(11.7)\qquad u_n'(0) = 1 \quad ,$$

for $n = 1,2,\ldots,N$.

In Eq. (11.7) we have chosen the eigenfunctions with unit initial slopes. Equations (11.4) through (11.7) pose a nonlinear two-point boundary-value problem, which we solve via quasilinearization.

12. Quasilinearization

Quasilinearization as a quadratically convergent successive

approximation technique which is well suited to modern digital computers, is described in Chapter II. We apply this method to a simple illustrative example. The results are given in Sec. 13.

Consider the case in which the form of k^2 is taken to be

$$(12.1)\qquad k^2(a,b;x) = 1 + ax + bx^2 .$$

The two eigenvalues

$$(12.2)\qquad \lambda_n = 1 - (n\pi)^2 , \quad n = 1,2,$$

are given. Numerically, these are $\lambda_1 = -8.8696043$ and $\lambda_2 = -38.478417$. We wish to estimate a and b.

We write the system of differential equations

$$(12.3)\qquad u_n'' = (\lambda_n - 1 - ax - bx^2)u_n , \quad n = 1,2,$$

$$(12.4)\qquad a' = 0 ,$$

$$(12.5)\qquad b' = 0 ,$$

in which a and b are viewed as dependent variables with zero derivatives. The initial conditions $u_n(0)$ and $u_n'(0)$ are given by Eqs. (11.6) and (11.7), but the initial conditions $a(0)$ and $b(0)$ are unknown. Let a^0 and b^0 be our initial estimates of a and b, and let $w_1(x) \approx u_1(x)$ and $w_2(x) \approx u_2(x)$ be our estimates of the eigenfunctions, produced by integrating Eqs. (12.3), (12.4) and (12.5) with initial values of Eqs. (11.6) and (11.7) and $a(0) = a^0$, $b(0) = b^0$.

The right-hand sides of Eqs. (12.3), (12.4) and (12.5) are expressed as expansions around the current approximations $w_1(x)$, $w_2(x)$, a^0 and b^0. Let the quantities in the new approximation be labeled $U_1(x)$, $U_2(x)$, A, and B. Then we have the system of linear differential equations

$$U_n'' = (\lambda_n - 1 - a^0x - b^0x^2)w_n$$

$$+ (U_n - w_n)(\lambda_n - 1 - a^0x - b^0x^2) + (A - a^0)(- xw_n)$$

$$(12.6) \qquad + (B - b^0)(- x^2w_n) , \qquad n = 1,2,$$

$$(12.7) \qquad A' = 0 ,$$

$$(12.8) \qquad B' = 0 .$$

The initial conditions are

$$(12.9) \qquad U_n(0) = 0 ,$$

$$(12.10) \qquad U_n'(0) = 1 , \qquad n = 1,2.$$

Let

$$(12.11) \qquad U_n(x) = p_n(x) + Aq_n(x) + Br_n(x) ,$$

$$n = 1,2,$$

$$(12.12) \qquad U_n'(x) = p_n'(x) + Aq_n'(x) + Br_n'(x) ,$$

where $p_n(x)$, $q_n(x)$ and $r_n(x)$ are solutions of

$$(12.13) \qquad p_n'' = (\lambda_n - 1 - a^0x - b^0x^2)p_n + (a^0x + b^0x^2)w_n , \qquad p_n(0) = 0$$

$$(12.14) \qquad q'' = (\lambda_n - 1 - a^0x - b^0x^2)q_n - x^2w_n , \qquad p_n'(0) = 0$$

$$q_n(0) = 0 , \qquad q_n'(0) = 0 ,$$

$$(12.15) \qquad r_n'' = (\lambda - 1 - a^0x - b^0x^2)r_n - x^2w_n ,$$

$$r_n(0) = 0 , \qquad r_n'(0) = 0 .$$

We produce $p_n(x)$, $q_n(x)$, and $r_n(x)$ by numerical integration. Note that initial conditions (12.9) and (12.10) are satisfied for any choice of A and B.

The constants A and B must be chosen so as to minimize S^*, where

$$(12.16) \quad S^* = \sum_{n=1}^{2} [p_n(1) + Aq_n(1) + Br_n(1)]^2 .$$

Thus, we require that

$$\frac{\partial S^*}{\partial A} = 0 \quad \text{and} \quad \frac{\partial S^*}{\partial B} = 0 ,$$

conditions which are equivalent to a system of two linear algebraic equations in two unknowns A and B. The solution of this system results in a new estimate of the differential operator and new approximations of the eigenfunctions. The procedure is repeated for a second approximation, beginning with Eq. (12.11). The function $U_n(x)$ then represents the second approximation, $w_n(x)$ the first approximation, etc. The procedure may be repeated again and again. In practice, only a few iterations are needed due to the rapid convergence of this technique. Of course, if the initial estimation is too poor, there may be no convergence.

13. Computational Results

As can be deduced from the eigenvalues in Eq. (12.2), the correct values of a and b are

$$(13.1) \quad a = 0 , \quad b = 0 .$$

Beginning with the initial estimates $a \approx 0.12$, $b \approx -0.23$, we obtain the results indicated in Table VI.

The program used an Adams-Moulton numerical integration subroutine with a step size of 0.0025. It takes only a few seconds to execute, and the results illustrate the rapid convergence which is to be expected. It should be borne in mind that the integration method is only producing

TABLE VI. Approximations of a and b

Approximation	a	b
0	0.12	-0.23
1	0.000002	-0.000455
2	-0.000027	+0.000049

4-5 significant figures and that the zeroes following the decimal point are significant.

14. Discussion

In experimental investigation, scanning techniques have proven to be useful in observing the behavior of a system under study. Yet if a more detailed understanding of the system is to be sought, then additional information must be extracted from the scan in a well defined manner.

The understanding of any physical system presupposes that a mathematical model of the system has been rationally formulated. Usually the mathematical model is structured as a set of defining differential equations, a set of system parameters and the associated initial and boundary conditions. A detailed understanding of a physical system, therefore, consists of knowing all the components of the mathematical model.

The methods we have explored in this chapter have allowed us to construct the initial conditions associated with a mathematical model governed by a partial differential equation where the system parameters are only those determining the initial conditions. Of particular interest is the speed and accuracy with which the initial conditions can be found. Using a nonlinear partial differential equation as an example, the success of the method can be judged from the results shown in Tables I-III. In each of the examples, the theoretical initial conditions dictated entirely different nonlinear solutions arising from the same differential equation. After four iterations each study shows good agreement with the exact initial values.

Based on these preliminary results we feel the method can be used to extract initial conditions from scanning data when the governing equation of the mathematical model is known.

We have also considered a problem where some of the eigenvalues are known. Again, numerical experiments indicate that the method is useful.

BIBLIOGRAPHY AND COMMENTS

1. Section 1. For differential quadrature, see R. Bellman, Methods of Nonlinear Analysis, Vol. I, (Academic Press, Inc., New York, 1969).

2. Section 9. See, for example,
   i. R. Bellman, Introduction to the Mathematical Theory of Control Processes, Vol. I, (Academic Press, Inc., New York, 1968).
   ii. R. Bellman, Introduction to the Mathematical Theory of Control Processes, Vol. II, (Academic Press, Inc., New York, 1971).

3. Some of the many interesting analytic problems which arise in the introduction of a new drug are discussed in the book by R. Bellman and C.P. Smith, Simulation in Human Systems - Decision-Making in Psychotherapy, (John Wiley & Sons, Inc., New York, 1973).

4. Section 12. We are following the paper by R. Bellman, H. Kagiwada and R. Kalaba, "Quasilinearization and the Estimation of Differential Operators from Eigenvalues", Comm. Assoc. for Computing Machinery, 11, No. 4, 255-256 (1968).

5. The work of Borg cited in Chapter XIII shows that we need a knowledge of two eigenvalue problems to determine the coefficient as pointed out there as his results do not give any quantitative estimates. We know that if the first n moments of a probability distribution are known we have inequalities for the probability distribution. We can expect that if we know the first n eigenvalues that there will be analogous inequalities for the coefficient function. However, no work has been done on this problem.

Chapter XIII

STRUCTURAL IDENTIFIABILITY

1. Introduction

A goal in many identification problems, in both industry and the biosciences, is to combine *a priori* knowledge with experimental data. By doing this we are frequently led to a model of the following type,

$$\frac{dx}{dt} = f(x,\theta,u) \ , \qquad y = g(x,\theta,u) \ , \tag{1.1}$$

where u is the input, y the output and x the state of the system. The vector θ denotes a set of unknown parameters. The purpose of the identification is to determine these parameters. One way to achieve this is to select an input signal u, measure the corresponding output signal y_m, and determine the parameters θ in such a way that the criterion

$$V(\theta) = \int_0^1 (y(t) - y_m(t))^2 \, dt \tag{1.2}$$

is minimal. Identification of mathematical models for industrial processes and compartmental analysis are typical examples of this approach.

In this chapter we will investigate if it is at all possible to determine the parameters θ using such a procedure. We shall then consider some general remarks about the procedure for a linear system. We shall begin by examining two examples.

Example 1. Consider the model,

$$\frac{dx}{dt} = ax + bu , \quad y = cx , \tag{1.3}$$

where $x, u,$ and y are unknown.

Assuming that the initial state of the system is known, we find that the input-output relations of the system can be written as

$$y(t) = bc \int_0^t \exp(a(t-s))\, u(s)ds \quad . \tag{1.4}$$

Since the parameters b and c enter into the input-output relation only as the product, bc, it is immediately clear that it is not possible to determine all the parameters a, b and c from the input-output relations. In order to get a unique solution, it is necessary to specify a or b or give a relation between them.

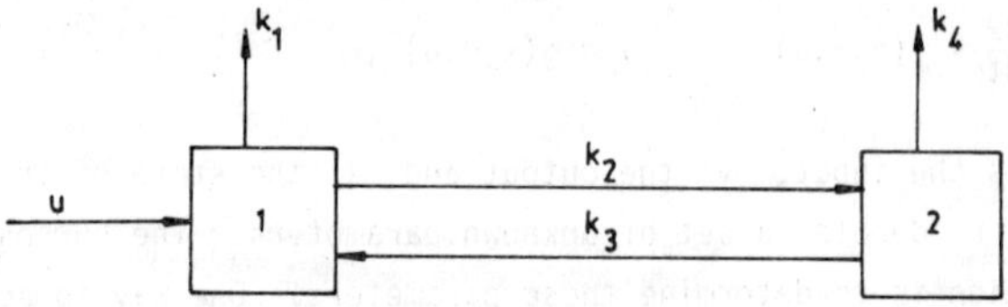

Fig. 13.1 Compartmental Model

Example 2. Consider the compartmental model shown in Fig. 13.1. Assume that the system is analysed by injecting a tracer into compartment 1 and taking samples from the same compartment. It will be shown that it is not possible to determine all the rate constants k_1, k_2, k_3 and k_4 from the data obtained in this way.

Let x_1 and x_2 be the concentrations of compartments 1 and

2. The equations describing the compartmental models are then

$$\frac{dx_1}{dt} = -(k_1 + k_2)x_1 + k_3x_2 + u ,$$

$$\frac{dx_2}{dt} = k_2x_1 - (k_3 + k_4)x_2 ,$$

$$y = x_1 . \tag{1.5}$$

By taking Laplace transforms we find that the input-output relation of the system (1.5) can be described by the transfer function

$$G(s) = \frac{(s + k_3 + k_4)}{\left(s^2 + s(k_1 + k_2 + k_3 + k_4) + (k_1 + k_2)(k_3 + k_4) - k_2k_3\right)} \tag{1.6}$$

We thus find that any parameter combination for which

$$k_3 + k_4 = \text{const} = b \tag{1.7}$$

$$k_1 + k_2 + k_3 + k_4 = \text{const} = a_1 \tag{1.8}$$

$$(k_1 + k_2)(k_3 + k_4) - k_2k_3 = \text{const} = a_2 \tag{1.9}$$

will give the same input-output relation. Since (1.7), (1.8), (1.9) give three equations to determine the four constants k_1, k_2, k_3, and k_4, we find immediately that the rate coefficients cannot be determined uniquely from input-output measurements. A one-parameter family of rate constants, which has the same input-output relation, can be constructed as follows: Select one coefficient arbitrarily. Solve Eqs. (1.7-1.9) for the other parameters.

Summing up, we find that it is not possible to determine all rate constants of the compartmental model of Fig. 13.1 from the data obtained by injecting a tracer in vessel 1 and taking samples from the same vessel.

Having realized that the problem is not trivial, let us proceed. The concept of identifiability is introduced in Sec. 2. A special case of identifiability of linear systems is discussed in Sec. 3. The following sections are devoted to special identifiable structures such as diagonal structures, companion structures and structures where all states are directly measurable. Compartmental models are discussed in Sec. 4.

2. Definitions

Consider a class of system S described by (1.1) where $\theta \varepsilon t$. By giving the functions f and g, we thus specify the internal coupling of the system as well as the coupling between the states and the inputs and outputs. For this reason we call S a structure. A structure can typically arise from a compartmental model of the type illustrated in Fig. 13.2, where some rate coefficients are specified and others unknown.

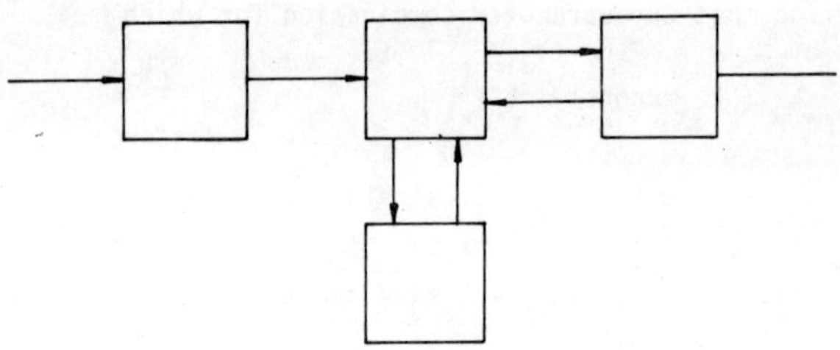

Fig. 13.2 A Typical Compartmental Structure

Efficient numerical methods for minimizing loss functions are well known. In this section we pose the problem whether for a given structure S it is possible to find a unique minimum at all. If this were the case, we could, for example, in the case of linear systems let f and g be arbitrary linear functions. By making experiments and solving the identification problem, we would get the coefficients of the matrices characterizing the linear functions and thus also the internal structure of the system. Unfortunately, this is not possible to do.

Compare examples 1 and 2. In order to get a unique solution, it is necessary to introduce restrictions.

The possibility of finding a unique solution to the identification problem depends not only on the structure of the system but also on the input signal. It is, however, possible to choose universal input signals which can be used for a large class of structures. For linear systems, we can choose the delta function. We will now specify what is meant by an identifiable structure.

Definition. Assume that the measured output is generated by a system S_0 with parameters θ_0. The structure S is (locally) identifiable if the function $V(\theta)$ has a local minimum at $\theta = \theta_0$. If the minimum is global, the structure is said to be globally identifiable. We immediately get the following criterion.

A sufficient condition for a structure S to be identifiable is that the matrix of the second-order partial derivatives with respect to the parameters, $V_{\theta\theta}$, is positive definite for all $\theta \varepsilon t$.

Remark 1. Notice that with regard to the practical applications, it is very important to define identifiability for a class of systems and not for a single system, since we will in practice never know that particular value of θ *a priori*. If we do, the identification problem would already be solved.

Remark 2. It is possible to formulate the identification problem in a more complex setting using probabilistic arguments. In a typical case the loss function V can then be taken as the negative logarithm of the likelihood function. In such a situation the expected value of $V_{\theta\theta}$ is equal to the information matrix of the estimation problem and the condition that $V_{\theta\theta}$ is positive definite can be given nice probabilistic interpretations.

It is naturally highly desirable to use identifiable structures when formulating and solving identification problems. If this is not the case, the parameters cannot be determined uniquely. If we attempt to identify a structure that is not identifiable, we will also encounter numerical problems. This is particularly the case in connection with algorithms like the Newton-Raphson method, which make use of the inverse of the matrix $V_{\theta\theta}$.

3. Linear Structures

We now consider linear structures, that is, a class of systems S whose elements $S = S(A,B,C)$, are described by the equations,

$$\frac{dx}{dt} = Ax + Bu \ , \qquad y = Cx \ , \tag{3.1}$$

where u is the input, y is the output and x is the state. It is assumed that the state vector is of dimension n. The matrices A, B, and C are assumed to have constant coefficients and $A \in \mathcal{A}$, $B \in \mathcal{B}$, and $C \in \mathcal{C}$, where $\mathcal{A}$, $\mathcal{B}$, and $\mathcal{C}$ are classes of matrices. The class $\mathcal{A}$ may be all 2×2 matrices with a_{11} positive, $a_{12} = 0$ and a_{21} and a_{22} arbitrary. By defining the sets A, B, and C we are thus specifying all internal couplings of the system.

We will now discuss the identifiability of the linear structure S characterized by (3.1). We first observe that since the input-output relation of (3.1) is uniquely determined by the controllable and observable parts of (3.1) it is necessary to assume that the system (3.1) is observable and controllable in the sense of Kalman for all $A \in \mathcal{A}$, $B \in \mathcal{B}$, and $C \in \mathcal{C}$.

Also assuming that the system is initially at rest, that is $\dot{x}(0) = 0$, the input-output relation is uniquely given by the impulse response of the system. From this we can immediately conclude that if it is at all possible to identify the linear structure, the identification can be achieved from an impulse response measurement. For a linear system it is thus possible to find at least one input signal, a delta function, that can be used universally. Admittedly, if there are disturbances, there might be other signals that are better, but we will not be concerned with that at this stage. We can also make another observation, namely, that all structures whose parameters are uniquely given by the impulse responses are identifiable. This probes the heuristic arguments used in the introduction. After these general observations, we now discuss some specific linear structures.

4. The Diagonal Structure

Consider a system in diagonal form, that is

$$(4.1)\qquad \frac{dx}{dt} = \begin{bmatrix} \lambda_1 & 0 & \cdots & 0 \\ 0 & \lambda_2 & \cdots & 0 \\ \cdot & \cdot & & \cdot \\ \cdot & \cdot & & \cdot \\ \cdot & \cdot & & \cdot \\ 0 & 0 & \cdots & \lambda_n \end{bmatrix} x + \begin{bmatrix} \beta_1 \\ \beta_2 \\ \cdot \\ \cdot \\ \cdot \\ \beta_n \end{bmatrix} u$$

The transfer function characterizing the input-output relation of the system is,

$$(4.2)\qquad G(s) = \sum_{i=1}^{n} \frac{B_i}{s + \lambda_i} \quad .$$

If all the characteristic values, λ's, are distinct, we find that the parameters B_i and λ_i can be uniquely determined from the impulse response. We can thus conclude that a diagonal structure with distinct characteristic values is identifiable. We can also prove this directly by showing that the matrix $V_{\theta\theta}$ is positive definite.

5. The Companion Structure

Consider a system in companion form, that is

$$(5.1)\qquad \frac{dx}{dt} = \begin{bmatrix} -a_1 & 1 & 0 & \cdots & 0 \\ -a_2 & 0 & 1 & \cdots & 0 \\ \cdot & \cdot & \cdot & & \cdot \\ \cdot & \cdot & \cdot & & \cdot \\ \cdot & \cdot & \cdot & & \cdot \\ -a_{n-1} & 0 & 0 & \cdots & 1 \\ -a_n & 0 & 0 & \cdots & 0 \end{bmatrix} x + \begin{bmatrix} b_1 \\ b_2 \\ \cdot \\ \cdot \\ \cdot \\ b_{n-1} \\ b_n \end{bmatrix} u ,$$

$$y = (1\ 0\ 0\ \dots\ 0)x, \qquad t \geq 0 .$$

Assuming that the initial state is zero, we find that the input-output relation of the system is given by

$$(5.2)\qquad \frac{Y(s)}{U(s)} = \frac{b_1 s^{n-1} + b_2 s^{n-2} + \dots + b_n}{s^n + a_1 s^{n-1} + \dots + a_n} ,$$

where Y and U are the Laplace transforms of y and u. If the polynomials

$$(5.3)\qquad A(s) = s^n + a_1 s^{n-1} + \dots + a_n$$

and

$$(5.4)\qquad B(s) = b_1 s^{n-1} + b_2 s^{n-2} + \dots + b_n$$

do not have a common factor, we find that the parameters a_i and b_i are uniquely given by the transfer function and so is the impulse response. We can thus conclude that the companion structure is identifiable.

Since the impulse response is invariant under the transformations

$$(5.5)\qquad A^* = TAT^{-1} , \qquad B^* = TB , \qquad C^* = CT^{-1}$$

we can also conclude that a structure $S+(A^\star, B^\star, C^\star)$ where $A^\star, B^\star, C^\star$ are given by (5.5), T is given and A, B, and C correspond to a controllable system in companion form, is identifiable.

6. A Linear Structure where all States are directly Observable

We now consider the particular case of a linear system (3.1) when the matrix C is of full rank. Without losing generality, we can assume that C equals the identity matrix. If this were not the case we could always introduce another output $y' = Cy$. The impulse response of the system is

$$h(t) = C e^{At} B = e^{At} B \quad . \tag{6.1}$$

We now show that if the system is completely controllable in Kalman's sense, that is, the matrix $(B, AB, \ldots, A^{n-1}B)$ has rank n, then the corresponding structure is identifiable. To do this we simply have to show that the matrices A and B can be determined uniquely from the impulse response. We have

$$B = h(0) \quad . \tag{6.2}$$

To determine A we observe that

$$h'(0) = AB = Ah(0) \quad ,$$

$$h''(0) = A^2B = Ah'(0) \; ,$$

$$h^{(n)}(0) = A^nB = Ah^{(n-1)}(0) \quad .$$

This follows directly by differentiating (6.1). Hence

$$A(h(0), h'(0), \ldots, h^{(n-1)}(0)) = (h'(0), h''(0), \ldots, h^{(n)}(0))$$

Since the system was assumed to be controllable, the matrix

$$(h(0), h'(0), \ldots, h^{(n-1)}(0)) = (B, AB, \ldots, A^{(n-1)}B)$$

is of rank n. Hence

$$A = \left(h'(0),\, h''(0), \ldots, h^{(n)}(0)\right)\left(h(0),\, h'(0), \ldots, h^{(n-1)}(0)\right)^{-1},$$

and we have thus shown how to determine A and B uniquely from the impulse response.

7. Some Compartmental Structures

We now investigate the identifiability of some compartmental structures. Consider the system with n cascaded compartments shown in Fig. 13.3.

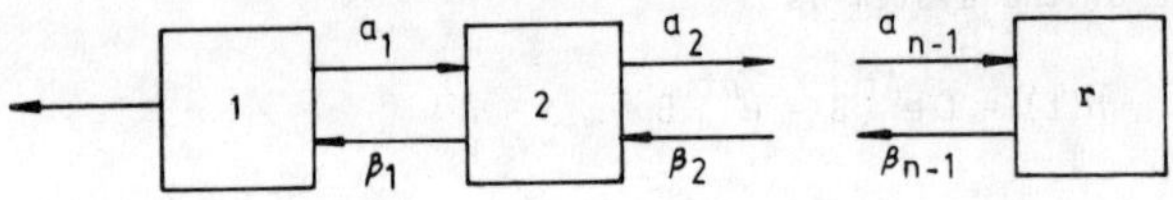

Fig. 13.3

Assume that the system is analyzed by injecting a tracer into compartment n and the concentration of the n-th compartment is considered as output. The system can be described by the following equations

$$(7.1)\quad \frac{dx}{dt} = \begin{bmatrix} -\alpha_1 & \beta_1 & 0 & \cdots & 0 & 0 \\ \alpha_1 & -(\alpha_2+\beta_1) & \beta_1 & \cdots & 0 & 0 \\ 0 & \alpha_2 & -(\alpha_3+\beta_2) & \cdots & 0 & 0 \\ \cdot & & & & & \cdot \\ \cdot & & & \cdots & & \cdot \\ \cdot & & & & & \cdot \\ 0 & 0 & 0 & \cdots & -(\alpha_{n-1}+\beta_{n-2}) & \beta_{n-1} \\ 0 & 0 & 0 & \cdots & \alpha_{n-1} & -\beta_{n-1} \end{bmatrix} x - \begin{bmatrix} 0 \\ 0 \\ 0 \\ \cdot \\ \cdot \\ \cdot \\ 0 \\ 1 \end{bmatrix} u,$$

$$y = (0\ 0\ 0\ \ldots\ 0\ 1)x \quad .$$

The transfer function of this system is given by

$$(7.2) \qquad G(s) = \frac{d_{n-1}(s)}{d_n(s)}$$

where

$$(7.3) \qquad d_n(s) = \det \begin{bmatrix} s+\alpha_1 & -\beta_1 & \cdots & 0 & 0 \\ -\alpha_1 & s+\alpha_2+\beta_1 & \cdots & 0 & 0 \\ \cdot & & & \cdot & \cdot \\ \cdot & & & \cdot & \cdot \\ \cdot & & & \cdot & \cdot \\ 0 & 0 & \cdots & s+\alpha_{n-1}+\beta_{n-2} & -\beta_{n-1} \\ 0 & 0 & \cdots & -\alpha_{n-1} & s+\alpha_n+\beta_{n-1} \end{bmatrix} .$$

To obtain a symmetrical form, we have introduced $\alpha_n = 0$. Using elementary rules for determinates, we get

$$(7.4) \qquad d_1(s) = s + \alpha_1$$

$$d_2(s) = (s + \alpha_2 + \beta_1)(s + \alpha_1) - \alpha_1\beta_1 \quad ,$$

$$\cdot$$

$$(7.5) \qquad \cdot$$

$$\cdot$$

$$d_k(s) = (s + \alpha_k + \beta_{k-1})d_{k-1}(s) - d_{k-1}\, \beta_{k-1}\, d_{k-2}(s) \quad .$$

The determinates can thus be evaluated recursively. It follows from Eqs. (7.2) and (7.4) that the parameters α_1 and β_1 can be determined uniquely from the transfer function. Putting $k = n$, we find from Eq. (7.4) that β_{n-1} and $\alpha_{n-1}\, \beta_{n-1}$ can be determined. Proceeding in this way we can determine all the other parameters. Notice in particular that we are not using the equation

$$d_1(s) = s + \alpha_1$$

which means that it would in fact be possible to have an additional parameter corresponding to the compartmental model of Fig. 13.4.

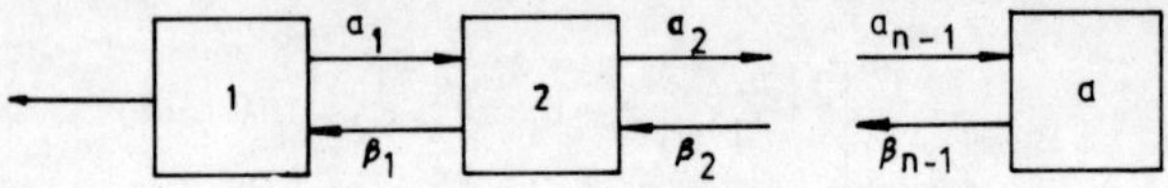

Fig. 13.4

Summing up, we find that the compartment models of Figs. 13.3 and 13.4 are identifiable if the tracer is injected into compartment a and if measurements are taken from the same compartment.

It is thus possible to determine all the rate constants of the models of Figs. 13.3 and 13.4 by a simple experiment, which consists of injecting a tracer into the n-th compartment and taking measurements from the same compartment.

Notice, however, that even for the compartmental model of Fig. 13.3, it is in general not possible to determine all the rate constants from an experiment arranged in a different way. This is illustrated by the following counter-example.

Consider a compartmental model according to Fig. 13.3 with three compartments. Assume the tracer is injected into the third compartment and that the system is analysed by taking samples from the first compartment. The equations governing the system then become

$$\text{(7.6)} \qquad \frac{dx}{dt} = \begin{bmatrix} -\alpha_1 & \beta_2 & 0 \\ \alpha_1 & -(\alpha_2+\beta_1) & \beta_2 \\ 0 & \alpha_2 & -\beta_2 \end{bmatrix} x + \begin{bmatrix} 0 \\ 0 \\ 1 \end{bmatrix} u \ ,$$

$$y = (1\ 0\ 0) \ .$$

The input-output relation of the linear system is characterized by the transfer function

$$G(s) = \frac{\beta_1\beta_2}{s^3 + s^2(\alpha_1 + \alpha_2 + \beta_1 + \beta_2) + s(\alpha_1\alpha_2 + \alpha_1\beta_2 + \beta_1\beta_2)} \quad (7.7)$$

The impulse response measurement or any other identification experiment with the same inputs and outputs give the following equations:

$$\beta_1\beta_2 = c_1 \ ,$$

$$\alpha_1 + \alpha_2 + \beta_1 + \beta_2 = c_2 \ ,$$

$$\alpha_1\alpha_2 + \alpha_1\beta_2 + \beta_1\beta_2 = c_3 \ . \quad (7.8)$$

It is obvious that it is impossible to determine the coefficients uniquely from these equations.

Also notice that if we have a compartmental model according to Fig. 13.3 with additional coupling (say a direct coupling between compartment 1 and compartment n-1), the corresponding structure will not be identifiable simply for the reason that there are more parameters in the system than the 2n parameters which characterize the impulse response.

Notice, however, that if measurements are taken from all compartments, it then follows from the results of Sec. 6 that if the tracer injection is chosen in such a way that the system is controllable, then the corresponding structure will be identifiable no matter how complex the interactions are.

8. Linear Systems with Constant Coefficients

In this section we will study the work of Walter and Cardinal who study the identifiability of linear state systems with constant coefficients, when some elements of the transition matrix can be measured on a finite interval of time. The results obtained by Walter et al. constitute an illuminating generalization of the ideas presented in the last sections.

We shall consider the transition matrix solution to the system:

$$\dot{X} = G\,X(t)\ , \quad X(0) = I\ . \tag{8.1}$$

It is easily shown that $X(t)$ satisfies the semi-group property,

$$X(t + t_o) = X(t)\,X(t_o)\ . \tag{8.2}$$

We quickly note that at $t_o = 0$, the identity (8.2) is true, further

$$\begin{aligned} \dot{X}(t + t_o) &= G\,X(t)\,X(t_o) \\ &= G\,X(t + t_o)\ . \end{aligned} \tag{8.3}$$

Assuming the eigenvalues of X are distinct, the matrix X can be written in the form

$$X(t) = \sum_{i=1}^{N} A_i\, e^{\alpha_i t}\ . \tag{8.4}$$

The matrix G is directly related to the set of solution matrices A_i by noting that from Eqs. (8.1) and (8.4),

$$\dot{X}(0) = \sum_{i=1}^{N} \alpha_i A_i = G\,X(0) = G\ . \tag{8.5}$$

An additional property is found by considering the expression,

$$\begin{aligned} X(t + t_o) &= \sum_{i=1}^{N} A_i\, e^{\alpha_i (t+t_o)} \\ &= \left(\sum_{i=1}^{N} A_i\, e^{\alpha_i t} \right) \left(\sum_{k=1}^{N} A_k\, e^{\alpha_k t} \right) \\ &= \sum_{i=1}^{N} \sum_{k=1}^{N} A_i\, A_k\, e^{(\alpha_i t + \alpha_k t_o)} \end{aligned} \tag{8.6}$$

For the identity above to hold,

$$A_i A_k = \delta_{ik} \quad . \tag{8.7}$$

Now since

$$G = \sum_{i=1}^{N} \alpha_i A_i \qquad ,$$

$$GA_j = \sum_{i=1}^{N} \alpha_i A_i A_j \quad ,$$

$$GA_j = \alpha_j A_j \quad .$$

The column vectors of A_j which are different from zero are then the eigenvectors of G with the same eigenvalues α_j. With the hypothesis of distinct eigenvalues α_k of G, the eigenvectors v_j of G are linearly independent and form a basis in the state space. Consequently all the eigenvectors of G having the same eigenvalue must be co-linear with v_j. Then all the column vectors of G are co-linear, and

$$\text{rank } (A_j) = 1 \quad . \tag{8.8}$$

The same demonstration would show that the row vectors of A_j are eigenvectors of G^T with the same eigenvalue α_j.

9. Identifiability Conditions

If the linear system (8.1) is completely characterized by the matrix G_o, then G_o represents a single point in the $n \times n$ parametric space associated with the problem. Suppose that a set of m pulse responses are known experimentally, that is m elements $x_{ij}(t)$ of the transition matrix X can be determined analytically without error. By definition we shall say that the matrix G is identifiable if the one and only one point $G_m = G_o$ of the parametric space leads to the analytic expressions of the measured elements of $x_{ij}(t)$.

Suppose that experimental curves are known in the analytic form

$$(9.1) \qquad x_{ij}(t) = \sum_{k=1}^{N} a_{ij}^{k} e^{\alpha_k t} .$$

For simplicity's sake, we shall assume that all the modes of the system are excited for the known elements of the transition matrix $x_{ij}(t)$, or

$$(9.2) \qquad a_{ij}^{k} \neq 0 , \qquad k = 1,\dots,N .$$

If the system (9.1) is of N-th order, the parametric space contains N^2 independent parameters g_{ij}; but the maximum number of independent parameters which can be deduced from a set of m observations of pulse responses amounts to

$$(9.3) \qquad N + m(N-1)$$

i.e., N parameters

nm parameters $a_{ij}^{k_k}$ which in view of (8.1) are related by

$$(9.4) \qquad x_{ij}(0) = \delta_{ij} = \sum_{k=1}^{N} a_{ij}^{k} .$$

It will be impossible to determine all the parameters if m is such that

$$(9.5) \qquad N + (N-1)m < N^2 ,$$

or

$$(9.6) \qquad N > m ;$$

thus a necessary identifiability condition is that at least N elements of X have been observed.

Now suppose that we know analytically a column of X. We know the N eigenvalues i of X, and N independent eigenvectors v_i of G, which are formed by the known columns of $A_i (i=1,\dots,N)$. These known vectors are the columns of a basis-conversion matrix P^{-1} which diagonalizes the G matrix. G can then be uniquely determined by,

$$(9.7) \qquad G = P^{-1}\, B\, P \quad ,$$

where

$$b_{ij} = \alpha_i\, \delta_{ij} \quad .$$

In the same way, the analytical knowledge of a row of X enables us to determine uniquely G^T and then G.

A sufficient identifiability condition is that a row or a column of X has been observed.

We have seen that the knowledge of a row or column of X enables this matrix to be uniquely determined and the corresponding differential system to be identified. Therefore if the property (8.8)

$$\text{rank}\ (A_j) = 1 \quad ,$$

enables a row or a column of X to be evaluated from the known elements of X, the system is identifiable. Using the fact that all second-rank determinates are equal to zero, the following conditions can be given:

a) A sufficient condition of identifiability is that $n + j - 1$ elements of X have been observed, located on j rows in such a way that

   (i) no known element can be deducted from the others using the property (8.8),

   (ii) no column of X is totally unknown.

b) Another sufficient condition can be obtained by exchanging the words row and column in the preceding condition.

c) If the necessary or the sufficient condition is not conclusive, the A_i matrices must be reconstructed from the known elements with respect to the relations (8.7) and (8.8). If all the coefficients are then determined uniquely, the system is identifiable. If, on the contrary, some undetermined parameters remain, the system is not identifiable. However, in a great many situations, the introduction of constraints on G may eliminate the indeterminacy when (8.5) is taken into account.

BIBLIOGRAPHY AND COMMENTS

1. Section 1. We are following the paper by R. Bellman and K. Astrom, "On Structural Identifiability", Math. Biosci., 7, Nos. 3 and 4, 329-330 (1970).

2. The term "structure" is a very useful one but it is very difficult to make it precise. It is clear that much topological and algebraic work is required.

3. In this connection let us mention the tearing technique of Kron which is found in the book by H.H. Happ, Diakoptics and Networks, (Academic Press, Inc., New York, 1970).

Chapter XIV

FINDING AN INITIAL APPROXIMATION

1. Introduction

In previous work, we have applied the theory of quasilinearization to various aspects of the identification of systems. By this last term we mean the determination of some properties of a system on the basis of observations of other properties. A typical problem of this genre is that of using a set of values $\{u(t_i)\}$, $i = 1,2,\ldots,M$, to ascertain the vector parameter a in the scalar differential equation

$$(1.1) \qquad u' = g(u,a) , \qquad u(0) = c .$$

Here $a = [a_1,a_2,\ldots,a_R]$. Often the initial condition, c, is also unknown.

The method hinges on the use of successive approximations. Let $a^{(0)}$ be an initial guess and let the function $u^{(0)}$ be determined by the equation

$$(1.2) \qquad u_0' = g(u_0, a^{(0)}) , \qquad u_0(0) = c .$$

Consider the linearized version of (1.1),

(1.3) $$u_1' = g(u_0, a_0^{(0)}) + g_u(u_1 - u_0) + J_a(a^{(1)} - a^{(0)}) \quad ,$$

where the partial derivative g_u and the Jacobian matrix J_a are evaluated at $u = u_0$, $a = a^{(0)}$. The vector $a^{(1)}$, and thus u_1, is determined by the condition that the expression

(1.4) $$D_M(a^{(1)}) = \sum_{i=1}^{M} (u_1(t_i) - b_i)^2$$

be minimized. Here b_i is the i-th observation, the quantity $u(t_i)$. Since $D_M(a^{(1)})$ is quadratic in the components of $a^{(1)}$, this minimization can be accomplished by means of the solution of a linear system of algebraic equations, or by a search technique.

This process is now repeated, obtaining in this fashion a sequence of vectors $\{a^{(k)}\}$ and functions $\{u_k(t)\}$. The convergence is quadratic, when convergence occurs. In general, we have convergence only if the initial guess $a^{(0)}$ is sufficiently close to the vector a.

In many cases this restriction is a severe drawback to the success of the method. It is interesting then to see if we can replace this qualitative condition by one of a quantitative nature.

2. Assumption of Polynomial Approximation

Instead of imposing the condition that $a^{(0)}$ be sufficiently close to a, let us impose a requirement on the analytic nature of $u(t)$. Specifically, we suppose that $u(t)$ is sufficiently smooth that the derivative at each t_i can be well approximated by a combination of the functional values at the entire set of points $t_1, t_2, \ldots, t_N$. The most convenient representation of this type is linear

(2.1) $$u'(t_i) = \sum_{j=1}^{M} a_{ij} u(t_j) \quad , \qquad i = 1,2,\ldots,M \quad .$$

There are several ways of prescribing the coefficients, as discussed below.

3. Determination of System Parameters

In a number of important cases the function $g(u,a)$ has the form

$$(3.1) \qquad g(u,a) = \sum_{k=1}^{N} a_k h_k(u) ,$$

an expression linear in the components of a. If we use the approximation of (2.1) we can replace (1.1) by the relations

$$(3.2) \qquad \sum_{j=1}^{M} a_{ij} u(t_j) = \sum_{k=1}^{N} a_k h_k(u(t_i)) .$$

The a_k may then be determined by a least squares procedure, namely, by the minimization of

$$(3.3) \qquad \sum_{i=1}^{M} \left[\sum_{j=1}^{M} a_{ij} u(t_j) - \sum_{k=1}^{N} a_k h_k(u(t_i)) \right]^2 .$$

4. Discussion

The foregoing procedure possesses two advantages over the quasilinearization approach of the introduction. It uses minimal qualitative assumptions concerning $u(t)$, and it requires no solution of functional equations. Furthermore, it can be used to provide an initial approximation $a^{(0)}$ if we do not wish to have a very accurate determination of a.

The approach can readily be extended to more general equations. If $g(u,a)$ is nonlinear in a, various approximation methods can be employed. Furthermore, we can readily modify it to handle the case where c is also unknown. The approximate representation of (2.1) can be extended to apply to differential-difference equations such as

$$(4.1) \qquad u'(t) + a_1 u(t-1) + a_2 u(t) = 0 ,$$

and to partial differential equations such as

$$u_{xx} + u_{yy} = g(u,a) \quad ,$$

$$(4.2) \qquad u_i - u_{jj} = g(u,a) \quad .$$

Furthermore, we can consider segmental approximations.

5. Determination of a_{ij}

We can determine the a_{ij} by the condition that the relation in (2.1) is valid for polynomials of degree less than or equal to M. This is most easily done by means of the orthogonal polynomials associated with the t_i. Alternatively, we can obtain a least squares polynomial fit of $u(t)$ and then calculate $u'(t_i)$.

BIBLIOGRAPHY AND COMMENTS

Section 1. Gauss was asked why he had used the square estimate of the error. He replied, "Why not?"

Chapter XV

DETERMINING A CRITERION

1. Introduction

We recall from our discussion of the quasilinear technique (Chapter II) that the device which allowed us to compute the system, initial conditions and parameters was the requirement that,

$$(1.1) \qquad \phi = \int_0^T \left[r(t) - p(t) + \sum_{i=1}^{n} c_i h_i(t))\right]^2 dt$$

be a minimum at each stage of the quasilinear iteration.

This well known technique is the method of least squares and, in this case, it determines the correct weighting of each of the homogeneous functions $h_i(t)$ in order to best fit the observed data. Yet this is not the only criterion which may be chosen in any given situation. This chapter will be devoted to the examination of the principle of the least squares and several sets of weighting functions, which could be used to improve the criterion in any particular situation.

At the end of the chapter we say a few words about the question of determining whether an observed behavior is optimal with respect to some criterion.

2. The Principle of Least Squares

The situation which is defined for the system identification problem is one in which the data $r(t)$ is known, presumably by observation, continuously in an interval $(0,T)$ and there are n independent functions $h_i(t)$, known also over the same interval, for which a least squared fit is to be made.

We then require that the approximation

$$r(t) = \sum_{k=0}^{n} c_k h_k(t) = y(t) \tag{2.1}$$

be as accurate as possible.

Suppose that a weighting function $w(t)$ is non-negative in the interval $(0,T)$,

$$w(t) \geq 0 \quad . \tag{2.2}$$

We now define the residual $R(t)$ by the equation,

$$R(t) = r(t) - \sum_{k=0}^{n} c_k h_k(t) = r(t) - y(t) \quad . \tag{2.3}$$

The best approximation, in the least squares sense, is defined to be that for which the c's are determined so that the integral of $w(t)R(t)$ is as small as possible. The requirement

$$\phi = \int_0^T w(t)\left[r(t) - \sum_{k=0}^{n} c_k h_k(t)\right]^2 dt \quad \text{is a minimum} \tag{2.4}$$

then imposes the conditions

$$\frac{\partial}{\partial c_r}\int_0^T w(t)\left[r(t) - \sum_{k=0}^{N} c_k h_k(t)\right]^2 dt = 0 \ , \qquad r = 0,1,\ldots,n \tag{2.5}$$

or

$$\sum_{k=0}^{n} c_k \int_0^T w(t)h_r(t)h_r(t)\, dt = \int_0^T w(t)h_r(t)r(t)\, dt \quad , \tag{2.6}$$

$$r = 0,1,\ldots,n \quad .$$

The system of equations given by (2.6) represents a linear system of algebraic equations of the general form,

$$D\,c = h \tag{2.7}$$

where D is an $n \times n$ symmetric matrix whose typical element is

$$d_{ij} = \int_0^T w(t)\; h_i(t)h_j(t)\, dt \quad , \tag{2.8}$$

$$c = \begin{Bmatrix} c_1 \\ \cdot \\ \cdot \\ c_n \end{Bmatrix} , \tag{2.9}$$ and h is the forcing vector whose i-th component is given by

$$h_i = \int_0^T w(t)\; h_i(t)\, dt \quad . \tag{2.10}$$

The solution of Eq. (2.7) is

$$c = D^{-1}\, h \quad . \tag{2.11}$$

The vector solution (2.11) is the vector of coefficients, which will minimize the square of the residual (2.4). It should be pointed out quickly that the solution (2.11) depends on the particular choice of the weighting function $w(t)$ defined over the interval of interest $(0,T)$.

3. <u>Uniform Weighting Function</u>

The most widely used method of determining a criterion is to assume that the errors incurred in using the data are uniformly distributed throughout the time interval. In such a case the weighting function is chosen as unity,

$$(3.1) \qquad w(t) = 1 , \quad 0 \le t \le T .$$

In this case,

$$(3.2) \qquad \phi = \int_0^T \left(r(t) - \sum_{k=1}^{n} c_k h_k(t) \right)^2 dt .$$

The requirement that ϕ must be a minimum leads to the set of algebraic equations,

$$(3.3) \qquad D\,c = h$$

where

$$(3.4) \qquad d_{ij} = \int_0^T h_i(t)\, h_j(t)\, dt ,$$

and

$$(3.5) \qquad h_i = \int_0^T r(t)\, h_i(t)\, dt .$$

4. <u>Laguerre Approximation</u>

In the case where the observed data is taken over a fairly long time interval and the initial conditions must be emphasized in the system identification procedure, it is useful to have a weighting function of the form,

$$(4.1) \qquad w(t) = e^{-at} ,$$

where a is a positive constant.

We must have,

$$(4.2) \qquad \phi = \int_0^T e^{-at} h_i(t) h_j(t)\, dt$$

and

$$(4.3) \qquad h_i = \int_0^T e^{-at} r(t) h_i(t)\, dt \quad .$$

5. Jacobi Approximation

In cases where errors occurring near the ends of an interval (a,b) are of particular importance, a weighting function, which is of the form $\frac{1}{(t-a)(b-t)}$, is often useful. For our set of problems $a = 0$, $b = T$, we have as a weighting function

$$(5.1) \qquad w(t) = \frac{1}{t(T-t)} \quad .$$

In this application,

$$(5.2) \qquad \phi = \int_0^T \frac{1}{t(T-t)} \left(r(t) - \sum_{k=1}^{n} c_k h_k(t) \right)^2 dt \quad .$$

The coefficients of the minimizing set of algebraic equations, $Dc = h$ are

$$(5.3) \qquad d_{ij} = \int_0^T \frac{1}{t(T-t)} h_i(t) h_j(t)\, dt$$

and

$$(5.4) \qquad h_i = \int_0^T \frac{1}{t(T-t)} r(t) h_i(t)\, dt \quad .$$

6. When is an Observed Behavior Optimal?

Let us mention a class of problems which we shall not pursue.

Often, we observe a given behavior and would like to know if that behavior is optimal with respect to some criterion. Procedures of this type are very valuable in physics. They yield a very simple way of deriving equations in a number of cases.

The reference given at the end of this chapter verges on the inverse problem of physics, one of the fundamental experimental problems.

BIBLIOGRAPHY AND COMMENTS

Section 6. Refer to the paper by R. Bellman, "Dynamic Programming and Inverse optimal Problems in Mathematical Economics", *J. of Math. Anal. Appl.*, **29**, No. 2, 424-428 (1970).

Chapter XVI

SUMS OF EXPONENTIALS

1. Introduction

So far in this book, we have considered the identification problem only within the context of a system of ordinary, and in some cases partial differential equations. One of the identification problems was defined as seeking the initial conditions and the system parameters associated with the set of governing equations.

There is, however, a wide body of problems which dominate the areas of mathematical biology, linear dynamics and systems with built-in delays in which the basic solution can be written as the sums of exponentials,

$$x(t) = c_1 e^{-a_1 t} + c_2 e^{-a_2 t} + \ldots + c_n d^{-a_n t} \quad . \tag{1.1}$$

It is particularly evident in the modeling of biological phenomena that the governing equations are of this form. Lotka, in his book "Elements of Mathematical Biology" defines the fundamental equations of kinetics of evolving systems to be of the form,

$$\frac{dx_i(t)}{dt} = f_i(x_1, x_2, \ldots, x_n; p) \; , \qquad i = 1,2,\ldots,n \tag{1.2}$$

where p is a set of system parameters. If this equation takes the form of a linear equation, i.e.

$$(1.3) \qquad \frac{dx_i(t)}{dt} = \sum_{j=1}^{n} a_{ij} x_j(t) \ ,$$

then the solution is expressible as a sum of exponentials.

In view of the observation that a great many mathematical systems are characterized by behavior given by Eq. (1.1), we shall devote this chapter to the examination of the origins of exponential solutions to system of equations and we shall discuss methods of identification in this particular situation.

While it will become clear that these particular systems, that is those admitting exponential solutions, can be identified using the techniques of quasilinearization, there are other methods which can be just as effective for the particular problems.

2. Linear Differential Equations

In the classical mathematical sense, an n-th other, ordinary, homogeneous, linear differential equation is defined as having the form,

$$(2.1) \qquad \frac{d^n x(t)}{dt^n} + \frac{a_{n-1}\, d^{n-1}\, x(t)}{dt^{n-1}} + \ldots + a_o x(t) = 0 \ .$$

If the coefficients a_k are constant, then the system (2.1) admits solutions which are sums of exponential for if we let,

$$(2.2) \qquad x(t) = C\, e^{\lambda t} \ ,$$

then by direct substitution into (2.1) we see that the differential equation is reduced to a polynomial,

$$(2.3) \qquad \lambda^n + a_{n-1}\, \lambda^{n-1} + \ldots + a_o = 0 \ .$$

The only solutions to (2.1) are those exponentials λ where the exponents are the roots of the characteristic polynomial (2.3). Because of the linearity of the system, the general solution to Eq. (1.1) can be written as

$$x(t) = c_1 e^{\lambda_1 t} f + c_2 e^{\lambda_2 t} + \ldots + c_n e^{\lambda_n t} \quad . \tag{2.4}$$

One can quickly see from the form of (2.4) that the exponential rate constants λ_i are a direct function of the system parameters a_i as they appear in the characteristic polynomial, and the multiplicative constants c_i reflect directly the initial conditions of the system.

A close examination of expression (2.4), shows that if we can identify both the constants c_i and the rate constants λ_i, then we have identified all the parameters we had previously found by the method of quasilinearization. The initial conditions of the system are directly given as the constants c_i, plus the rate constants,

$$x(0) = c_1 + c_2 + \ldots + c_n$$

$$\frac{dx(0)}{dt} = c_1\lambda_1 + c_2\lambda_2 + \ldots + c_n\lambda_n$$

$$\frac{d^2x(0)}{dt^2} = c_1\lambda_2^2 + c_2\lambda_2^2 + \ldots + c_n\lambda_n^2 \quad . \tag{2.5}$$

The system parameters a_i can be recovered directly from the rate constants by recalling that these constants are the roots of the system characteristic polynomial, so we can write,

$$(\lambda-\lambda_1)(\lambda-\lambda_2) \ldots (\lambda-\lambda_n) = 0 \quad . \tag{2.6}$$

By expanding, Eq. (2.6) can be written in the form of Eq. (2.3), thereby directly identifying the system parameters a_i.

What we have shown here is that the identification problem can be solved by a method very different from the quasilinear technique, thus adding more analytical tools to our arsenal.

3. Linear Differential-Difference Equations

In addition to systems governed by linear ordinary differential equations, sums of exponentials are also the solutions to differential-difference equations, which have many applications in biological modeling. By a differential-difference equation we shall mean an equation in an unknown function and certain of its derivatives, evaluated at arguments which differ by any of a fixed number of values. Examples of such equations are the equations,

$$\frac{d^2u}{dt^2}(t) - \frac{du}{dt}(t-2) + u(t) = 0 \quad ,$$

$$\frac{du}{dt}(t) - u(t-1) - u(t-2) = 0 \quad ,$$

$$\frac{du}{dt}(t) - 2u(t) + \frac{du}{dt}(t-1) - 2u(t-1) = e^{2t} \quad . \tag{3.1}$$

In this section we will be interested exclusively with linear equations of the form,

$$\sum_{i=0}^{m}\sum_{j=0}^{n} a_{ij}(t)\frac{d^j}{dt^j}(t-\omega_i) = f(t) \quad , \tag{3.2}$$

where m and n are positive integers, $0 = \omega_0 < \omega_1 < \dots < \omega_n$ and where $f(t)$ and the $(m+1)(n+1)$ functions $a_{ij}(t)$ are defined in some interval of the real t-axis.

Moreover, our primary emphasis will be upon linear equations with constant coefficients of the form,

$$\sum_{i=0}^{m}\sum_{j=0}^{n} a_{ij}\frac{d^j}{dt^j}(t-\omega_i) = f(t) \quad . \tag{3.3}$$

We shall further restrict our attention to the subclass of equations of the form (3.3) which are of first order in derivatives and differences. Such equations can be written in the form,

$$(3.4) \qquad a_0 \frac{du}{dt}(t) + a_1 \frac{du}{dt}(t-\omega) + b_0 u(t) + b_1 u(t-\omega) = f(t) \quad .$$

In this section we shall be concerned with examining methods for building solutions to (3.4) as a sum of simple exponential solutions. This method is, of course, well known in the theory of differential equations. It is the fundamental method of superposition on which mathematical physics rests.

For convenience of notation, we shall define a linear operator $L(u)$ by the following equation,

$$(3.5) \qquad L(u) + a_0 \frac{du}{dt}(t) + b_0 u(t) + b_1 u(t-\omega) \quad .$$

Due to the linearity of the operator $L(u)$, we can immediately make the following statements:

If $u_1(t)$ and $u_2(t)$ are any two solutions of the equation $L(u) = 0$, and if c_1 and c_2 are any two constants, then $c_1 u_1(t) + c_2 u_2(t)$ is also a solution of $L(u) = 0$.

We also observe that, if $v(t)$ is a solution of $L(u) = f$, and if $w(t)$ is a solution of $L(u) = 0$, then $v(t) + w(t)$ is a solution of $L(u) = f$.

This is clear, since $L(v + w) = L(v) + L(w) = f$.

The linearity of the operator $L(u)$ suggests the possibility of generating every solution of the homogeneous equation as a linear combination of simple solutions, as done in the theory of ordinary differential equations. Since in the latter theory the simple solutions are exponentials, it is not surprising that we can also find exponential solutions of differential-difference equations. Indeed if we assume a solution of the form,

$$(3.6) \qquad u(t) = e^{st} \quad ,$$

then substitution into Eq. (3.5) yields a transcendental, or characteristic equation,

$$(3.7) \qquad h(s) = a_o s + b_o + b_1 e^{-\omega s} .$$

Only for values of s for which $h(s) = 0$, do exponential solutions to the differential-difference equations exist.

The general results for differential-difference equations can be characterized in the following statement:

$$L(u) = a_o u'(t) + b_o u(t) + b_1 u(t-\omega) = 0$$

is satisfied by

$$\sum p_r(t)\, e^{s_r t} ,$$

where s_r is any sequence of characteristic roots of L, $p_r(t)$ is a polynomial of degree less than the multiplicity of s_r, and the sum is either finite or infinite with suitable conditions to ensure convergence.

4. Systems of Linear Differential Equations

While we have shown in Sec. 2 that a single linear homogeneous differential equation with constant coefficients gives rise to sums of exponentials as solutions, the same is also true for systems of homogeneous differential equations with constant coefficients. Consider the system

$$(4.1) \qquad \frac{dx_i(t)}{dt} = \sum_{j=1}^{N} a_{ij} x_j(t) , \qquad i = 1,2,\ldots,N .$$

This system of ordinary differential equations occur frequently in the modeling of systems, particularly biological, and therefore we have great interest in examining the possible solutions.

Indeed if we let,

$$(4.2) \qquad x_i(t) = b_i e^{\lambda t} ,$$

the direct substitution of expression (4.2) into Eq. (4.1) yields the following well known characteristic value problem,

$$\text{(4.3)} \qquad \begin{bmatrix} a_{11}-\lambda & a_{12} & \cdots & a_{1N} \\ a_{21} & a_{22}-\lambda & \cdots & a_{2N} \\ a_{N1} & a_{N2} & \cdots & a_{NN}-\lambda \end{bmatrix} \begin{bmatrix} b_1 \\ b_2 \\ b_N \end{bmatrix} = \begin{bmatrix} 0 \\ 0 \\ 0 \end{bmatrix}$$

The set of equations (4.3) yields a nontrival set of solutions only when the determinant,

$$\text{(4.4)} \qquad \begin{vmatrix} a_{11}-\lambda & a_{12} & \cdots & a_{1N} \\ a_{21} & a_{22}-\lambda & \cdots & a_{2N} \\ a_{N1} & a_{N2} & \cdots & a_{NN}-\lambda \end{vmatrix} = 0 \quad .$$

The condition that the determinant vanishes is equivalent to finding the zeroes of the characteristic polynomial,

$$\text{(4.5)} \qquad \lambda^N + c_1\lambda^{N-1} + \ldots + c_{N-1} = 0 \quad .$$

Once the zeroes of the characteristic polynomial are found, then for each distinct characteristic value λ_i, there exists a characteristic vector e_i so that the general solution to the system of Eqs. (4.1) have the form,

$$\text{(4.6)} \qquad x(t) = \bar{\alpha_i} e_1 + \bar{\alpha_2} e_2 + \ldots + \bar{\alpha_N} e_N \quad ,$$

again clearly showing the solution as a sum of exponentials.

5. Systems of Differential-Difference Equations

With the aid of the appropriate notation, it is fairly easy to extend the results of Sec. 3 to more general equations. We shall consider the linear system of equations,

$$(5.1)\qquad \sum_{i=0}^{m}\sum_{j=1}^{n} a_{ijk}u'_j(t-\omega_i) + \sum_{i=0}^{m}\sum_{j=1}^{n} b_{ijk}u_j(t-\omega_i) = f_k(t) \quad ,$$

$$k = 1,2,\ldots,n \quad ,$$

involving n unknown functions $u_1(t),\ldots,u_n(t)$ and their first derivatives.

With the aid of the vector-matrix notation, we shall write the general linear system of equations in (4.1) in simplified form. Let y and f represent the column vectors

$$(5.2)\qquad y = \begin{bmatrix} u_1 \\ u_2 \\ \cdot \\ \cdot \\ \cdot \\ u_n \end{bmatrix} \quad , \qquad f = \begin{bmatrix} f_1 \\ f_2 \\ \cdot \\ \cdot \\ \cdot \\ f_n \end{bmatrix} \quad .$$

Let A_i be the matrix of the k-th element which is a_{ijk} and let B_i be the matrix of the kj-th element which is b_{ijk}. We see at once that the system takes the form:

$$(5.3)\qquad \sum_{i=0}^{m} A_i y'(t-\omega_i) + \sum_{i=0}^{m} B_i y(t-\omega) = f(t) \quad .$$

The characteristic equation corresponding to this system is the transcendental equation,

$$(5.4)\qquad \det \sum_{i=0}^{m} (A_i s + B_i)\exp(-\omega_i s) = 0 \quad .$$

The set of solutions to Eq. (4.5) forms the basis for the construction of the solution to the system of linear differential-difference equations as again a sum of exponentials.

6. Compartmental Analysis

A large class of mathematical models which give rise to sums of exponentials as solutions are the linear compartmental models. An important application is the analysis of data obtained from tracer kinetic studies. In this particular analysis the system is represented as a model consisting of several phases or compartments. Then the primary objective of a tracer experiment is the determination of the number and size of the compartments and the rate constants between interacting compartments.

A linear compartmental system is usually specified by a number of compartments and by transition rates (probabilities) between compartments. The behavior within each compartment is represented by a linear ordinary differential equation,

$$\frac{d}{dx_i}(t) = \sum_{i=1}^{n} b_{ij} x_j(t) \quad , \qquad i = 1,2,\ldots,n \ . \tag{6.1}$$

When the b_{ij} matrix has distinct eigenvalues, the solutions of Eqs. (6.1), for the constant b_{ij}, are the sums of exponentials

$$x_i(t) = \sum_{j=1}^{n} B_{ij} e^{-c_j t} \tag{6.2}$$

where $-c_j$'s are the eigenvalues of the b_{ij} matrix and B_{ij}'s are the elements of the eigenvectors.

Explicit solutions for the case of multiple roots and some time-dependent b_{ij} may also be derived.

To derive a compartmental model, the solution which will agree with experimental data that are measures of $x(t)$, the number of compartments and the values of the rate constants, b_{ij}, must be determined. Strictly speaking there is no unique model for a set of data since one can always expand any model by the introduction of degeneracies or fine structure beyond the resolution possible from that data. In practice, however, one usually seeks a model with the minimum number of compartments, unless independent information is available to the contrary.

In setting up a model some *a priori* knowledge about the system may be available, but more frequently one has to be guided by the data. The most useful clue is the fact that the solution must be a linear combination of exponentials, to agree with the solutions of the differential equations and that the same set of exponentials must apply to all compartments (except for degeneracies). This suggests that the data be fitted first to a linear combination of exponentials and that the number of compartments for the model be chosen to equal the number of exponentials in the fit.

7. Prony's Method and the Identification of Parameters

So far this chapter has been devoted entirely to the study of mathematical systems which give rise to solutions in the form of sums of exponentials. If the solution to such a system has the form,

$$(7.1) \qquad x(t) = c_1 e^{a_1 t} + c_2 e^{a_2 t} + \ldots + c_n e^{a_n t}$$

then we have shown that a knowledge of the coefficients and the rate constants in (7.1) will allow us to reconstruct the initial conditions and the system parameters of the original problem. Therefore by identifying these two sets of parameters, we are in fact solving the identification problem.

To be sure, the set of equations governing the basic problem under study could be transformed, using quasilinearization, into a problem of systems identification for which we are familiar, yet it is to our advantage to consider other methods which would yield the same result. Such a method is Prony's Method which we shall now consider in detail.

Consider a system whose solution can be written in the form of Eq. (1.1). If we make the substitution,

$$(7.2) \qquad p_k = e^{a_k} \quad ,$$

then we can write,

(7.3) $$x(t) = c_1p_1^{\,t} + c_2p_2^{\,t} + \dots + c_np_n^{\,t}$$

We suppose that a linear change of variables has been introduced, in advance, in such a way that values of $x(t)$ are specified at N equally spaced points $t = 0,1,\dots,N-1$.

If the set of points are to fit the expression (7.3), then it must be true that,

$$c_1 + c_2 + \dots + c_n = x(0) \quad ,$$

$$c_1p_1 + c_2p_2 + \dots + c_np_n = x(1) \quad ,$$

$$c_1p_1^{\,2} + c_2p_2^{\,2} + \dots + c_np_n^{\,2} = x(2) \quad ,$$

(7.4) $$c_1p_1^{\,N-1} + c_2p_2^{\,N-1} + \dots + c_np_n^{\,N-1} = x(N-1) \quad .$$

If the constants $p_1,\dots,p_n$ are known, then this set comprises n linear equations in n unknowns and can be solved exactly. If $N > n$, then the unknown constants are found approximately by the least squares technique.

If, however, the p_k must also be determined, at least $2n$ equations are needed which are now nonlinear in the p's. This difficulty can be minimized by the method which we will describe next.

Let $p_1,\dots,p_n$ be the roots of the algebraic equation

(7.5) $$p^n - d_1p^{n-1} - d_2p^{n-2} - \dots - d_n = 0 \quad ,$$

so that the left-hand member of (7.5) is identified with the product $(p - p_1)(p - p_2)\dots(p - p_n)$. In order to determine the coefficients $d_1,\dots,d_n$, we multiply the first equation of (7.4) by d_n, the second by $d_{n-1},\dots,$ the n-th equation by d_1 and the (n+1)th equation by -1 and add the results. If we made use of the fact that each p satisfies (7.5), the result is seen to be of the form,

(7.6) $$x(n) - d_1x(n-1) - d_2x(n-2) - \dots = 0 \quad .$$

A set of N-n-1 equations can be set up in the same way by starting instead with the second, third,..., (N-n) equations. In this way we find that (7.4) and (7.5) imply the N-n linear equations;

$$x(n-1)d_1 + x(n-2)d_2 + \dots + x(0)d_n = x(n) \quad ,$$

$$x(n)d_1 + x(n-1)d_2 + \dots + x(1)d_n = x(n+1) \quad ,$$

$$x(N-2)d_1 + x(N-3)d_2 + \dots + x(N-n-1)d_n = x(N-1) \quad . \tag{7.7}$$

Since the ordinates $x(k)$ are known, this set can be solved directly for the n d's of $N=2n$ or solved approximately by the method of least squares if $N > 2n$.

After the d's are determined, the p's are found as the roots of (7.5). They may be real or complex. The equations (7.4) then become a set of linear equations in the coefficients c_i. The c's can be determined from any set of n equations or, preferably, by applying the least squares technique to the entire set.

Thus the nonlinearity of the system is concentrated in the single algebraic equation (7.5). The technique described is known as Prony's method.

8. Comparison of Prony's Method with the Method of Quasilinearization

Throughout this chapter we have observed that the discussion of these models arising from linear homogeneous differential equations, linear differential-difference equations and those from compartmental analysis, all form a set of first order linear differential equations and all have sums of exponentials as solutions.

Therefore, we have a choice in the selection of the method of system identification which may be employed in any particular problem. We can either consider the set of differential equations governing the problem as the basic set of equations to work with, or we may recognize the fact that the solution is a set of exponentials and seek to determine the coefficients of the exponential fit directly from the observed data.

In evaluating a particular problem, several competing facts must be kept in mind. If the quasilinearization technique is used, the solution yields the unknown information to the identification problem directly, namely the initial conditions and the system parameters. On the other hand, we choose to fit the data to a set of exponentials, then it is necessary to use that information to reconstruct both the initial conditions and the system parameters.

Numerically, the two methods present very different problems both in techniques of numerical solution and storage capacity. If one selects to solve the problem direclty using quasilinearization, then the procedure consists of solving a rather large set of quasilinear ordinary differential equations over a given time interval and storing the results of each integration. The data fitting technique then requires the minimization of an error measurement. This is reflected in a matrix inversion to determine the solution to the identification problem.

The method of exponential fits, using Prony's method, involves the inversion of a much larger system of algebraic equations, followed by the determination of the n roots of the characteristic polynomial. These roots may be complex making the numerical solution more difficult. Finally, the determination of the initial conditions and the system parameters, which are required for a solution to the identification problem, can be found directly from the information contained in the exponential data fit.

Basically one has to choose between a technique which used more storage, inverts smaller matrices and solves quasilinear equations, which the quasilinear technique does, and Prony's method which inverts a much larger matrix and must find the roots of a polynomial, which is a nonlinear numerical process.

9. Discussion

The solution to the systems identification problem is by no means limited to the techniques of quasilinearization. For that matter the ideas of fitting sums of exponentials to observed data is not unique. In fact scores of methods exist for fitting continuous functions to observed experimental data.

The proper interpretation of the various data fits, however, have long been lacking in so far as system identification is concerned. To fit a continuous curve to a set of observed data points really says little about the underlying structure of the system. A clue to the understanding of the deep structure of a system can be seen in the use of compartmental analysis to define the system. Once the data fitting technique is associated with a model, in most cases a set of differential equations, then an appreciation of the intricate interrelations between assumed states in the system can be made with some rational basis.

Thus data fitting to sums of exponential plus the associated compartmental model, shows a way of determining optimal models. With some knowledge of the optimal model, the quasilinear technique can be brought to bare in seeking a fuller understanding of the system under consideration.

10. On the Number of Exponentials

In a number of different fields, it is known as a result of an a priori theoretical analysis that a sequence of numbers $\{u_n\}$, obtained experimentally, has the representation

$$u_n = \sum_{k=1}^{N} c_k e^{\lambda_k n} \qquad (10.1)$$

If the dimension N is known, the parameters c_k and λ_k can be determined provided that enough of the u_n are given. If N is an unknown, as it is in many investigations where the determination of the number of components is one of the major objects, the problem is much more complex, and thus far more interesting.

The purpose of these sections is to outline a method designed to yield the value of N, without a simultaneous determination of the c_k and λ_k.

11. Recurrence Relations

We start from the frequently used fact that a sequence such as that given above can be generated by means of the recurrence relation, or difference equation,

(11.1) $$u_{t+N} = a_1 u_{t+N-1} + \dots + a_N u_t , \qquad t = 0,1,\dots .$$

with $u_0, u_1, \dots, u_{N-1}$ prescribed. Consider the determinant

(11.2) $$C_k(t) = \begin{vmatrix} u_t & u_{t+1} & \cdots & u_{t+k} \\ u_{t+1} & u_{t+2} & \cdots & u_{t+k+1} \\ \cdot & \cdot & \cdots & \cdot \\ \cdot & \cdot & \cdots & \cdot \\ \cdot & \cdot & \cdots & \cdot \\ u_{t+k} & u_{t+k+1} & \cdots & u_{t+2k} \end{vmatrix}$$

related to the CASORATI determinant.

We may assume that in Eq. (1.1) $c_1 \neq 0$, $i = 1,2,\dots,N$, and that $\lambda_i \neq \lambda_j$ for $i \neq j$.

It is then easy to see that

(11.3a) $C_k(t) = 0$, $k \geq N$, $t = 0,1,\dots$,

(11.3b) $C_{N-1}(t) = C_{N-1}(0)a_N^t \neq 0$, $t = 0,1,\dots$,

(11.3c) $C_k(t) \neq 0$, $k < N-1$, r large t .

It follows that the functions $C_k(t)$, $k = 1,2,\dots$, can be used as a sequence of test functions to determine N, given the numerical values of u_n.

12. Discussion

The most unsatisfactory criterion in (11.3) is (11.3c). In many cases this can be considerably strengthened, making use of the fact that the λ_k's are real and the C_k's are positive. In this case, it follows from the known criteria for the positive definiteness of quadratic forms, or from the equivalent known properties of Gramian determinants, that $C_k(t) > 0$, $k < N-1$.

BIBLIOGRAPHY AND COMMENTS

1. Section 2. It should be noted that exponentials also arise from linear difference equations.

2. Section 3. On the general theory of differential-difference equations, see R. Bellman and K. L. Cooke, *Differential-Difference Equations,* (Academic Press, Inc., New York, 1963).

3. Section 10. We are following the paper by R. Bellman, "On the Separation of Exponentials", *Boll. U.M.I.,* *15*, No. 3, 38-39 (1960).

Chapter XVII

FINDING TWO DEFECTIVE COINS

1. Introduction

In this chapter, we shall study some aspects of the problem of finding two defective coins among similar coins. This is a prototype of the problem of finding a fault in a complex system. We shall give one method of studying this problem, using the theory of dynamic programming.

2. Finding Defective Coins

The problem of ascertaining the minimum number of weighings which suffice to determine the defective coins in a set of N similar coins given an equal arm balance and the information that there is precisely one defective coin present, is well known. A large number of ingenious solutions exist, some based upon sequential procedures and some not.

Oddly enough, the corresponding problem for finding more than one defective coin has attracted little attention. At our suggestion this problem was investigated by Cairns. Apart from this unpublished paper, little appears to be written on the subject. The problem is of significance because it represents one of the simplest examples of a sequential testing problem replete with the difficulties of combinatorial nature and the difficulties inherent in the concept of "information".

A small amount of analysis discloses the enormous difference in complexity between the one-coin and two-coin problem. Starting with a set of N coins known to contain precisely one defective coin, and an equal arm balance, the only permissible operation is that of weighing one group of k coins against another group of k coins chosen from the original set. If the two groups balance, we immediately conclude that the defective coin is in the remaining group of $N-2k$ coins. If the two groups do not balance, we know that the defective coin is in one of the two sets of k coins. Taking the simple case where we know in advance that the defective coin is heavier than any of the other coins, we see that regardless of the outcome of the first test, we face a problem of exactly the same type as the original at the end of the test, and with a smaller number of coins. This invariance permits us to apply the theory of dynamic programming to this problem in a quite simple fashion.

The problem in the case where there are two or more defective coins is far more complex because we cannot draw any such definite conclusion at the end of a single test. We shall analyze this in detail in the following section. The purpose is to indicate a systematic way in which the theory of dynamic programming can be used to provide a computational solution to the determination of optimal and suboptimal testing policies. We shall illustrate this by means of some numerical results obtained using a digital computer.

Let us note in passing that an alternative approach has been given by Sobel and Groll in the investigation of a problem due to Dorfman.

3. The Two-Coin Problem

Let us now consider the case where there are two defective coins in a set of N coins, and where these two coins are known to be heavier than the other coins in the set. Let us consider the possibilities upon a single weighing involving two sets of k coins each.

(a) If the two sets balance, either they both possess one defective coin, with no defective coins remaining in the group of $N-2k$ coins, or each group of k possesses no defective coins and the two defectives are in the remaining group of $N-2k$ coins.

(b) If the two sets do not balance, either one set possesses one defective coin with at most one in the remaining $N-2k$ coins, or one set of k coins possesses two defectives with none in the remaining $N-2k$.

We see that the original information pattern has a ramification. As the testing continues, this ramification increases at an alarming rate if we allow all possible testing problems. Observe that the set of possible tests has become very much larger than what it was in the simple case where there is only one defective coin.

At the end of the first test, we can choose a subset from the first group of k coins, another subset from the second set of k coins, and a third subset from the group of $N-2k$ coins, and then match this set of coins against a similar set. Although many of these tests are clearly inefficient and majorized by simpler policies, it is not easy to rule out all tests of this type without a good deal of difficult analyses. Furthermore, it is not clear *a priori* that some tests of this nature may not be useful.

In what follows, however, in order to simplify the analysis and speed up the calculations, we shall not allow any mixed tests of the type described above. We shall allow the information obtained from the testing of one group of coins to be used in the testing of any other group.

4. Analytic Formulation

From what has been said above, we see that regardless of the definite information which we start with, and regardless of the outcome of a particular test, we end up in a situation in which we must face alternatives. Let us then assume from the very beginning that we are given an initial probability $p = [p_0, p_1, p_2]$, where p_i is the probability that there are exactly i defective coins in the set of N coins, for $i = 0,1,2$. Assuming that these defective coins are distributed in a uniform fashion among the N coins, we can immediately calculate the corresponding probability distribution for any set of k coins drawn from the original group of N.

We shall only consider policies of the following type: Given a set of N coins and an associated probability distribution $p = [p_0, p_1, p_2]$, we draw two groups of k coins each at random and balance them against one another. If they balance, which means that each group now has at most one defective coin, we search one of the groups. If this has no defective coin, then the other batch of k coins has none, and we go on to examine the remaining group of $N-2k$ coins. If the first batch has one defective coin, then the second batch has exactly one defective coin too, and there is no need to examine the remaining $N-2k$ coins.

If the two groups of k are unbalanced, we proceed to isolate the defective coins in the heavier batch. If it has one defective coin, then the other batch of k has no defective coin, and we proceed to find the remaining defective coin in the group of $N-2k$ coins, if there is indeed one more. If the k-group has two defective coins, we need not examine the other two groups of coins.

Prior to writing down the functional equation we shall use to determine the optimal choice of k, let us introduce the following quantities:

$f_N(p_0,p_1,p_2)$[a] = expected number of weighings required to determine all defective coins, given p_0,p_1,p_2, and using a suboptimal procedure for N coins

$q_N(p_0,p_1; k)$ = Prob (two k-batches balance)

$P_B(0)$ = Prob (no defective coins in each k-batch, given that the arms balance)

---

[a]This will also be referred to as $f_N(p_o,p_1)$, since $p_2 = 1-p_o-p_1$ and is hence redundant.

$P_B(1)$ = Prob (one defective coin in each, given that the arms balance)

$P_U(1)$ = Prob (one defective coin in one k-batch and none in the other, given that the arms are unbalanced)

$P_U(2)$ = Prob (two defective coins in one k-batch, s in the second k-batch, and t in the untested N-2k batch)

$p(r;s;t)$ = Prob (r defective coins in the first k-batch, s in the second k-batch, and t in the untested N-2k batch)

$f_N(0,p_1,p_2) \equiv g_N(0,p_1,p_2) \equiv g_N(p_1,p_2) \equiv g_N(p_1)$.

$f_N(p_0,p_1,0) \equiv h_N(p_0,p_1,0) \equiv h_N(p_0,p_1) \equiv h_N(p_0)$

$f_N(0,1,0) \equiv d_N(0,1,0) \equiv d_N$

5. Values of p(r;s;t)

In order to derive the basic recurrence equations, it will be useful to tabulate the various values of $p(r;s;t)$:

$$p(1;1;0) = \frac{2k^2}{N(N-1)} p_2 \quad ;$$

$$p(1;0;1) = \frac{2k(N-2k)}{N(N-1)} p_2 \quad ;$$

$$p(0;1;1) = \frac{k1(N-2k)}{N(N-1)} p_2 \quad ;$$

$$p(2;0;0) = \frac{k(k-1)}{N(N-1)} p_2 \quad ;$$

$$p(0;2;0) = \frac{k(k-1)}{N(N-1)} p_2 \quad ;$$

$$p(0;0;2) = \frac{(N-2k)(N-2k-1)}{N(N-1)} p_2 \quad ;$$

$$p(0;1;0) = \frac{k}{N} p_1 \quad ;$$

$$p(1;0;0) = \frac{k}{N}\, p_1 \quad ;$$

$$p(0;0;1) = \frac{N-2k}{N}\, p_0 \quad ;$$

(5.1) $$p(0;0;0) = p_0 \quad .$$

6. The Basic Functional Equation

Applying the principle of optimality in the usual fashion, the basic equation from which the search policies are derived is readily seen to be

$$\begin{aligned} f_N(p_0,p_1,p_2) = \min_k \Big\{ 1 + q_N(p_0,p_1;k)\Big[h_k(p_0^{(1)}) \\ + P_B(0) f_{N-2k}(p_0^{(2)},\ p_1^{(2)},\ p_2^{(2)}) + P_B(1)\, d_k\Big] \\ + (1-q_N(p_0;p_1;k))\Big[g_k(p_1^{(3)}) + P_U(1) h_{N-2k}(p_0^{(4)})\Big]\Big\} , \end{aligned} \tag{6.1}$$

where the $p_1^{(j)}$ are probabilities conditional upon the various possible results of the weighing of the two k-batches, and

(6.2) $$\sum_i p_1^{(j)} = 1 , \quad \text{all } j \quad .$$

The superscript (1) corresponds to a balance of the two groups of k. The superscript (2) corresponds to the case where the defective coins are known to be in the remaining $N-2k$ coins. The superscript (3) corresponds to an unbalance of the original two groups of k, and the superscript (4) to the penultimate case described in the previous section.

The $p_i^{(j)}$ are given by

$$p_0^{(1)} = \frac{p_0 + p_2\ [(N-2k)(N-2k-1)\ /\ N(N-1)]}{p_0 + p_2\big[(N-2k)(N-2k-1) + 2k^2\big]/\ N(N-1)} \quad ;$$

$$p_1^{(1)} = 1 - p_0^{(1)} \quad ;$$

$$p_0^{(2)} = \frac{p_0}{p_0 + p_1[(N-2k)/N] + p_2[(N-2k)(N-2k-1)/N(N-1)]} \quad ;$$

$$p_1^{(2)} = \frac{p_1(N-2k)/N}{p_0 + p_1[(N-2k)/N] + p_2[(N-2k)(N-2k-1)/N(N-1)]} \quad ;$$

$$p_2^{(2)} = 1 - p_0^{(2)} - p_1^{(2)} \quad ;$$

$$p_0^{(3)} = 0 \quad ;$$

$$p_1^{(3)} = \frac{p_1(2k/N) + p_2[4k(N-2k)/N(N-1)]}{p_1(2k/N) + p_2[4k(N-2k) + 2k(k-1)]/N(N-1)} \quad ;$$

$$p_2^{(3)} = \frac{p_2[2k(k-1)/N(N-1)]}{p_1(2k/N) + p_2[4k(N-2k) + 2k(k-1)]/N(N-1)} \quad ;$$

$$p_0^{(4)} = \frac{p(2k/N)}{p_1(2k/N) + p_2[4k(N-2k)/N(N-1)]} \quad ;$$

$$p_1^{(4)} = 1 - p_0^{(4)} \quad ;$$

(6.3) $$p_2^{(4)} = 0 \quad .$$

Also

(6.4) $$q_N(p_0,p_1;k) = p_0 + p_1\,\frac{(N-2k)}{N} + p_2\left[\frac{(N-2k)(N-2k-1) + 2k^2}{N(N-1)}\right] \quad ,$$

and

$$P_B(0) = p_0^{(1)} \quad ;$$

$$P_B(1) = p_1^{(1)} \quad ;$$

(6.5) $$P_U(1) = p_1\,\frac{2k}{N} + p_2\,\frac{4k(N-2k)}{N(N-1)}$$

The equations for d_N, $h_N(p_0)$, and $g_N(p_1)$ are special cases of Eq. (6.1), reduced to:

$$d_N = \min_k \left[1 + \frac{N-2k}{N} d_{N-2k} + \frac{2k}{N} d_k\right] ,$$

$$h_N(p_0) = \min_k \Big[1 + q_N(p_0,\, 1-p_0;\, k) h_{N-2k}(p_0^{(2)})$$

$$+ (1-q_k\,(p_0,\, 1-p_0;\, k))d_k\Big] , \qquad (6.6)$$

and

$$g_N(p_1) = \min_k \Big\{ 1 + q_N(0,p_1;k)\Big[h_k(p_0^{(1)}) + P_B(0)g_{N-2k}(p_1^{(2)})$$

$$+ P_B(1)d_k\Big] + (1-q_N(0,p_1;k))\Big[g_k(p_1^{(3)}$$

$$+ P_U(1)H_{N-2k}(p_0^{(4)})\Big]\Big\} . \qquad (6.7)$$

7. Computational Procedures

Using Eqs. (6.1), (6.6) and (6.7), approximate values of f_N and the corresponding values of k which yield the minimal were obtained for $N=1,\ldots,99$, for values of p_i at increments of 0.1. These values of f_N and d were computed iteratively using a UNIVAC 1105. Due to the existing storage size of computers today, to the dimensionality of the problem, and to the limited amount of computer time available, the method of computation was rather crude and hence the results give only an approximate idea as to the nature of the exact solution. However, some clues are obtained. It will be possible to obtain more accurate solutions using exactly the same techniques with smaller increments on the p, when computers with larger memories become available.

The crude results are due to the fact that the $p_i^{(j)}$ themselves take on values in the continuum $[0,1]$ for the various values of p_0, p_1, p_2, k, and N. That is, the $p_i^{(j)}$ calculated from Eqs. (6.3) are not

0, 0.1, 0.2,...,0.9, 1.0, but take on intermediary values whose corresponding functional values $f_r(p_0^{(j)}, p_1^{(j)}, p_2^{(j)})$'s have to be approximated by interpolation on previously obtained $f_r(p_0', p_1', p_2')$'s, where p_0', p_1', p_2' have values 0, 0.1,...,0.9, or 1.0.

We shall call these eleven values the set P; also $p_i^{(j)} > p_i^{-(j)}$ are those consecutive members of P that contain $p_i^{(j)}$.

For the cases in which $p_0^{-(j)} + p_1^{-(j)} \not> 1$, the interpolation on $f_r(p_0^{(j)}, p_1^{(j)})$ (where $p_1^{(j)}$ is redundant, since $p_0^{(j)} + p_1^{(j)} + p_2^{(j)} = 1$), is of the form

$$f_r(p_0,p_1) = 0.01(p_0-p_0)f_r(\bar{p}_0,\bar{p}_1) + (\bar{p}_0-p_0)f_r(p_0,p_1)(p_1-p_1)$$

(7.1) $$+ 0.01(p_0-p_0)f_r(\bar{p}_0,p_1) + (\bar{p}_0-p_0)f_r(p_0,p_1)(\bar{p}_1-p_1) .$$

When $p_0^{(j)} + p_1^{(j)} = 1.1$, the interpolation has to be modified so that only $f_r(p_0,p_1)$, $f_r(p_0,p_1)$, and $f_r(p_0,p_1)$ occur in the expression. Obviously, the incremental interval 0.1 is large and creates inaccuracies in the computations, but the fact that all $f_r(p_0,p_1)$, $r < N$, had to be stored to obtain f_N made a smaller interval impossible. Also, one would intuitively feel that the f's and their corresponding k's which minimize (which we shall define as k_N) might be discontinuous at the boundaries $p_0^{(j)} = 1$, $p_1^{(j)} = 1$, and $p_0^{(j)} + p_1^{(j)} = 1$, and this again is a drawback to the use of interpolation.

Since the number of values of $f_N(p_0,p_1)$ calculated is over 6000, limitations of space preclude printing all of them and their corresponding k_N's. However, as examples of the basic data obtained, Tables I to IV show these data for N = 80 and 90.

From the $f_N(p_0,p_1,p_2)$ and their corresponding k_N's obtained, various properties of the process that suggest themselves are:

(7.2a) $$f_N(0,0,1) = \log_2 N \quad ,$$

(7.2b) $$f_N(0,01) = \max_{|p_i|} f_N(p_0,p_1,p_2) \quad ,$$

(7.2c) $\dfrac{f_N(p_0,p_1,p_2)}{\log_2 N}$ tends to a limit as $N \to \infty$, p_0, p_1, p_2 remaining fixed.

The validity of the last statement is suggested by a statistical analysis performed on the data for $N = 79$ to 99; it was decided to use these values of N to make things as asymptotic as possible while still retaining sufficient data to provide indications of trends. For these values of N, and for all $\{p_i\}$ in P, the statistics

$$r_N(p_0,p_1) = \frac{k_N(p_0,p_1)}{N}$$

and

$$s_N(p_0,p_1) = \frac{f_N(p_0,p_1)}{\log_2 N}$$

TABLE I. Values of k_N for $N = 80$

p_n \ p_1	0	0.1	0.2	0.3	0.4	0.5	0.6	0.7	0.8	0.9	1.0
0	7	7	7	8	8	8	18	18	40	40	27
0.1	3	7	8	8	10	10	40	40	40	27	
0.2	4	10	10	12	14	40	40	40	29		
0.3	5	8	9	10	40	40	40	40			
0.4	7	7	40	40	40	40	40				
0.5	8	8	40	40	21	40					
0.6	7	38	40	40	40						
0.7	7	38	39	40							
0.8	32	37	39								
0.9	28	37									
1.0	—										

TABLE II. Values of f_N for N = 80

p_n \ p_1	0	0.1	0.2	0.3	0.4	0.5	0.6	0.7	0.8	0.9	1.0
0	6.447	6.512	6.539	6.535	6.497	6.416	6.283	6.072	5.748	5.143	1.000
0.1	4.674	5.773	6.146	6.309	6.298	6.161	5.928	5.496	4.913	3.927	
0.2	4.534	5.203	5.736	5.988	5.837	5.588	5.204	4.663	3.720		
0.3	4.344	4.920	5.422	5.486	5.213	4.911	4.408	3.450			
0.4	4.056	4.680	5.114	4.890	4.619	4.154	3.100				
0.5	3.781	4.340	4.524	4.272	3.835	2.750					
0.6	3.452	4.058	3.917	3.415	2.400						
0.7	3.160	3.488	3.021	2.050							
0.8	2.706	2.481	1.675								
0.9	1.715	1.315									
1.0	—										

TABLE III. Values of k_N for N = 90

p_n \ p_1	0	0.1	0.2	0.3	0.4	0.5	0.6	0.7	0.8	0.9	1.0
0	7	8	8	9	10	10	10	22	45	45	27
0.1	5	8	10	10	10	10	22	45	45	31	
0.2	5	10	10	10	14	15	45	45	32		
0.3	6	9	9	45	45	45	45	45			
0.4	6	8	9	45	45	45	45				
0.5	9	9	45	45	45	45					
0.6	8	10	44	45	45						
0.7	8	37	44	44							
0.8	36	42	44								
0.9	32	41									
1.0	—										

TABLE IV. Values of f_N for N = 90

p_n \ p_1	0	0.1	0.2	0.3	0.4	0.5	0.6	0.7	0.8	0.9	1.0
0	6.591	6.657	6.690	6.687	6.634	6.553	6.432	6.236	5.909	5.272	4.156
0.1	4.770	5.895	6.269	6.437	6.436	6.321	6.103	5.656	5.030	4.069	
0.2	4.664	5.409	5.810	6.085	5.943	5.694	5.343	4.765	3.849		
0.3	4.450	5.034	5.508	5.654	5.396	5.033	4.501	3.520			
0.4	4.153	4.767	5.208	5.004	4.733	4.238	3.160				
0.5	5.867	4.422	4.607	4.371	3.913	2.800					
0.6	5.523	4.134	3.988	3.513	2.440						
0.7	3.217	3.511	3.061	2.053							
0.8	2.747	2.527	1.702								
0.9	1.748	1.320									
1.0	—										

were computed. For fixed p_i, the $s_N(p_0,p_1)$ are amazingly constant over N, examples being given in Table V. Quadratic regressions on p_0 and p_1 were performed on the means of r_N and s_N (averaged over N); i.e., on the statistics

$$r(p_0,p_1) = \frac{1}{21}\sum_{N=79}^{99} r_N(p_0,p_1) \quad ,$$

$$s(p_0,p_1) = \frac{1}{21}\sum_{N=79}^{99} s_N(p_0,p_1) \quad . \tag{7.3}$$

The regressions are:

$$r = -1.076 + 6.014p_0 - 0.594p_0^2 + 8.350p_1 - 3.431p_1^2 \quad ,$$

$$\text{with } \sigma_r = 0.956 \quad ,$$

and

$$s = 1.017 - 0.869p_0 + 0.013p_0^2 + 0.388p_1 - 0.764p_1^2 \quad ,$$

$$\text{with } \sigma_s = 0.070 \quad .$$

Equation (7.5), with its low variance and its constant near unity, suggests that the regression is fairly accurate.

However, Eq. (7.4), with its high value of σ_r, is not too satisfactory. One reason for this result is that the k_N's are in some cases highly erratic through N for this fixed p_0, p_1. This is probably partly because (i) there might be some k's which produce almost the same value of f and the crudity of the computations does not prefer the best consistently, and (ii) the interpolations, as mentioned previously, might not give good results at the boundaries $p_0^{(j)} = 1$, $p_1^{(j)} = 1$, and $p_0^{(j)} + p_1^{(j)} = 1$. Reason (i) is backed up by the basic data in that for many values of p_0, p_1, two highly different values of k occur consistently and one of them is probably the correct one.

Table V

N	$S_N(0,0)$	$S_N(0.2, 0.5)$	$S_N(0.6, 0.3)$	$S_N(0.8, 0.1)$
79	1.470	1.305	0.784	0.570
80	1.471	1.275	0.786	0.566
81	1.460	1.290	0.779	0.571
82	1.470	1.282	0.788	0.565
83	1.468	1.292	0.781	0.570
84	1.468	1.262	0.785	0.564
85	1.466	1.291	0.778	0.568
86	1.467	1.268	0.786	0.562
87	1.464	1.277	0.778	0.566
88	1.465	1.271	0.779	0.562
89	1.462	1.281	0.772	0.566
90	1.465	1.265	0.781	0.561
91	1.463	1.289	0.774	0.563
92	1.464	1.274	0.777	0.561
93	1.462	1.283	0.770	0.557
94	1.463	1.275	0.779	0.559
95	1.461	1.283	0.773	0.555
96	1.462	1.268	0.773	0.558
97	1.460	1.281	0.767	0.556
98	1.461	1.270	0.776	0.557
99	1.460	1.278	0.770	0.553

Examples from the basic data appear in Table V.

8. Branching of Probabilities

The foregoing procedure illustrates very clearly how probabilities can multiply at an alarming rate. There are many processes where this phenomenon occurs. It would seem that in considering such processes we have to use some approximation techniques to keep the possibilities within bounds.

9. Discussion

The proper selection of a criterion to be associated with the quasilinearization technique is not an easy one for there are no hard and fast rules to follow.

BIBLIOGRAPHY AND COMMENTS

Section 1. The discussion here follows the paper by R. Bellman and B. Gluss, "On Various Versions of the Defective Coin Problem", *Information and Control*, 4, 118-131 (1969).

Chapter XVIII

THE MOTION OF A ROCKET IN A SMOOTHBORE LAUNCHER

1. Introduction

In this chapter quasilinearization is used to study the motion of a spinning rocket inside a smoothbore launcher. We include this example to illustrate how quasilinearization can be applied to a complex identification problem.

In this particular instance the initial impulse imparted to a rocket and the dynamic coefficient of friction between the rocket and the wall of the launcher had to be determined. From a practical standpoint the motion of the rocket can be observed upon exit from the launcher making the initial condition difficult to measure directly. By directly recording the motion of the rocket as it left the launch tube we seek to determine the unknown initial condition and system parameter.

2. The Model

A nonlinear model is derived to include both the forward and rotary motion of the rocket coupled through the dynamic coefficient of friction.

The physical dimensions of the system are assumed to be known, while the initial conditions and the dynamic coefficient of friction are unknown.

3. The Mathematical Problem

The launcher rocket assembly is assumed to be made up of a cylindrical launch tube of length ℓ_0, and radius r_0, plus the rocket approximated by a solid cylinder of length ℓ_1, radius r_1, and having mass m, shown in Fig. 18.1.

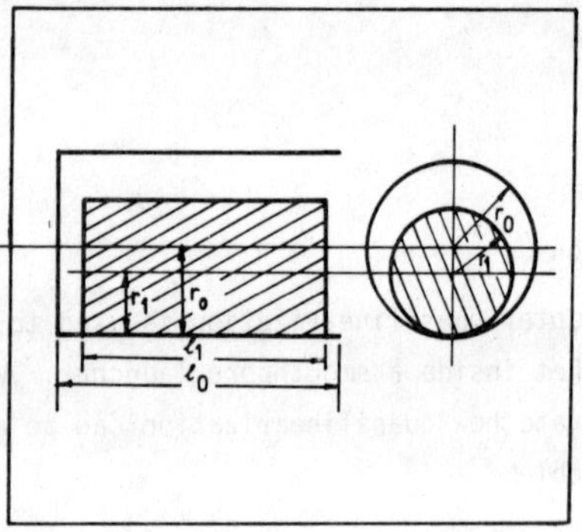

Fig. 18.1 Launcher Rocket Assembly

Figure 18.2 shows the dependent variables of the system, namely the motion of the contact point $\phi(t)$, the angular motion of the round, $\theta(t)$, and the forward motion, $x(t)$. The force system is shown in Fig. 18.3.

4. The Equations of Motion

The differential equations of motion are

$$\frac{m\delta d^2\phi}{dt^2} = - mg \sin \phi - F_\phi \quad ,$$

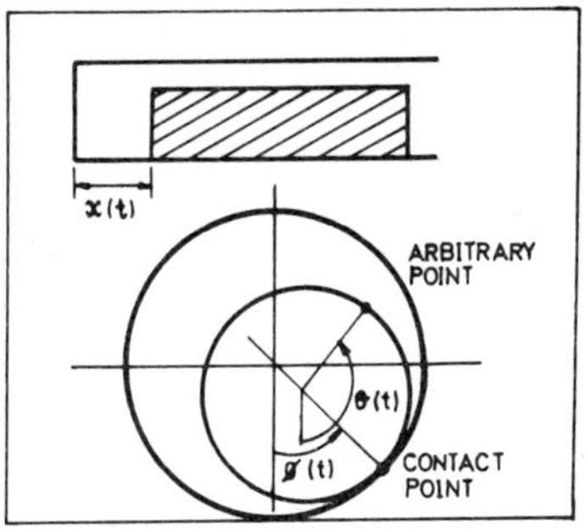

Fig. 18.2 Dependent Variables of the System

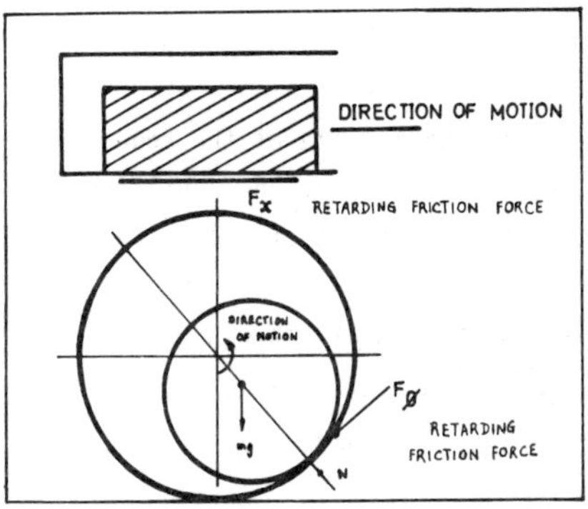

Fig. 18.3 The Force System

$$m\delta \left(\frac{d\phi}{dt}\right)^2 = -mg\cos\phi + N \quad ,$$

$$\frac{md^2x}{dt^2} = -F_x \quad ,$$

$$\frac{Id^2\theta}{dt^2} = -r_1F_\phi \quad , \tag{4.1}$$

while the initial conditions are

$$\phi(0) = c_1 \quad ,$$

$$\theta(0) = c_2 \quad ,$$

$$x(0) = c_3 \quad ,$$

$$\frac{d\phi(0)}{dt} = c_4 \quad ,$$

$$\frac{d\theta(0)}{dt} = c_5 \quad ,$$

$$\frac{dx(0)}{dt} = c_6 \quad , \tag{4.2}$$

where

I is the moment of inertia of the round,

N is the reacting force at the contact point,

$\delta = r_0 - r_1$ is the offset of the rocket in the bore.

5. The Non-Dimensional Equations of Motion

The following non-dimensional parameters and variables are introduced into the system by defining

$$\xi = \frac{x}{r_1} \quad , \qquad \rho = \frac{N}{w} \quad ,$$

$$\varepsilon = \frac{r_0}{r_1} - 1.0 \quad , \qquad f_\phi = \frac{F_\phi}{w} \quad ,$$

$$\tau = t\sqrt{\frac{g}{r_1}} \quad , \qquad f_x = \frac{F_x}{w} \quad , \tag{5.1}$$

where w is the weight of the rocket.

The non-dimensional equations of motion can be written

$$\varepsilon \frac{d^2\phi}{d\tau^2} + \sin\phi + f_\phi = 0 \quad , \qquad \phi(0) = c_1 : \frac{d\phi(0)}{d\tau} = c_4 \quad ,$$

$$\varepsilon\left(\frac{d\phi}{d\tau}\right)^2 + \cos\phi = \rho \quad , \qquad \theta(0) = c_2 : \frac{d\theta(0)}{d\tau} = c_5 \quad ,$$

$$\frac{d^2\theta}{d\tau^2} + 2f_\phi = 0 \quad , \qquad \xi(0) = c_3 : \frac{d\xi(0)}{d\tau} = c_6 \quad ,$$

(5.2) $$\frac{d^2\xi}{d\tau^2} + f_x = 0 \quad ,$$

where for the solid rocket (considered to be a disk), $I = \frac{1}{2} mr_1$. The reaction forces of friction f_ϕ and f_x are considered to be components of a single friction force vector f whose magnitude is proportional to the reaction force ρ and whose direction is defined as β. These are defined below:

$$f_\phi = f \sin\beta$$

$$f_x = f \cos\beta$$

$$f = \nu\rho$$

$$\tan\beta = \frac{\varepsilon\, d\phi/d\tau + d\theta/d\tau}{d\xi/d\tau}$$

To simplify the equations further let

$$\tau = s\sqrt{\varepsilon}$$

$$\theta = \varepsilon\Omega \ , \qquad (\ )' = \frac{d}{ds}$$

$$\xi = \varepsilon\eta \quad .$$

The equations of motion then have the following form:

$$\phi'' + \sin\phi + f_\phi = 0$$

(5.3) $$(\phi')^2 + \cos\phi = \rho\ , \qquad \Omega'' + 2f_\phi = 0\ , \qquad \eta'' + f_\eta = 0\ ,$$

where

$$f_\phi = \nu_1\rho\sin\beta\ , \qquad f_\eta = \nu_1\rho\cos\beta\ ,$$

$$\beta = \tan^{-1}\left(\frac{\phi' + \Omega'}{\eta'}\right)\ .$$

The parameter ν represents the dynamic coefficient of friction. A new variable $A(s)$ can be introduced into the system by letting

$$\phi' + \Omega' = A\sin\beta\ , \qquad \eta' = A\cos\beta\ .$$

Then the variables $\Omega(s)$ and $\eta(s)$ can be eliminated from the system giving

$$\phi'' + \sin\phi + \nu\rho\sin\beta = 0\ , \qquad (\phi')^2 + \cos\phi = \rho$$

$$A'\sin\beta + (A\cos\beta)\beta' + 2\nu\rho\sin\beta - \phi'' = 0$$

(5.4) $$A'\cos\beta - (A\sin\beta)\beta' + \nu\rho\cos\beta = 0\ .$$

Then ρ can be eliminated, and the final non-dimensional equations of motion can be written in the following form:

$$\phi'' = -\sin\phi - \nu[(\phi')^2 + \cos\phi]\sin\beta$$

$$A' = -\sin\phi\sin\beta - \nu[(\phi')^2 + \cos\phi](1 + 2\sin^2\beta)$$

(5.5) $$\beta' = -\frac{1}{A}\left([\sin\phi + 2\nu\{(\phi')^2 + \cos\phi\}\sin\beta]\cos\beta\right)\ .$$

The initial conditions of this set of equations can be derived directly from the set of initial conditions given in Eqs. (5.1).

6. The Inverse Problem

The set of differential equations, Eq. (5.5), defines the motion of a rocket through a smoothbore launcher. If the value of the friction coefficient ν, plus the initial conditions are known, then the solution of Eq. (5.5) may be found by straightforward numerical techniques.

Consider the inverse problem. If the values of $\phi(t)$, $\theta(t)$, and $x(t)$ over a definite time interval, are known from experimental observation, then can the initial conditions and the coefficient of friction, which will define a solution of Eq. (5.5) and which will give a "best" least square fit to the observed data over the entire time interval be estimated. Clearly such information would be useful in this particular problem. The coefficient of friction, which represents the complex retarding force characteristics on the round as it moves through the bore, is unknown. Since the round is initially totally within the bore chamber, the initial conditions are also not well known.

Such qualitative engineering problems as the determination of the initial impulses imparted to the round by the rocket motors and the evaluation of the effect of friction on the motion of the round through the launch tube, can be studied by closer examination of the solution to the inverse problem stated above.

7. The First-Order Differential Equations

Since the complete solution of the problem is desired, namely, knowledge of $\phi(t)$, $\theta(t)$, and $x(t)$, consider the following set of equations, written in the non-dimensional time variable, s.

(7.1a) $\phi' = \gamma$

(7.1b) $\theta' = \varepsilon(A \sin \beta - \gamma)$

(7.1c) $x' = r_1 \varepsilon A \cos \beta$

(7.1d) $\gamma' = -\sin \phi - \nu[\gamma^2 + \cos \phi] \sin \beta$

(7.1e) $A' = -\sin \phi \sin \beta - \nu[\gamma^2 + \cos \phi](1 + 2 \sin^2 \beta)$

(7.1f) $\beta' = \frac{-1}{A} [\sin \phi + 2\nu[\gamma^2 + \cos \phi] \sin \beta] \cos \beta$.

The initial conditions of Eqs. (7.1a - 7.1f) can be expressed in terms of the initial conditions of Eqs. (4.2).

8. Quasilinear Equations

Following the quasilinear technique, the set of Eqs. (7.1a - 7.1f), when written in vector form becomes

$$\frac{dZ_i^{n+1}}{dt} = f_i(Z^n) + \sum_{j=1}^{7} \frac{\partial f_j(Z^n)}{\partial Z_i}\left(Z_i^{n+1} - Z_i^n\right), \qquad i = 1,\ldots 7 .$$

The vector component Z_7 corresponds to the unknown parameter ν,
$\frac{dZ_7}{dt} = 0, \ Z_7(0) = \nu$.

9. The Observed Data

Let the quantities $\phi(t)$, $\theta(t)$ and $x(t)$ be observed over a time interval $[0,T]$. In non-dimensional terms, let them be $\xi_1(\tau)$, $\xi_2(\tau)$ and $\xi_3(\tau)$ respectively.

Then the error vector ψ can be introduced with components,

$$\psi_i(\tau) = \xi_i(\tau) - z_i^{n+1}(\tau) , \qquad i = 1,2,3$$

or

$$\psi_i(\tau) = \xi_i(\tau) - \left(\rho(\tau) + \sum_{k=1}^{7} {}_k^{n+1} (\tau)\right) , \qquad i = 1,2,3 .$$

The solution will be required to give the best least square fit to the observed data over the time span $(0,T)$, that is,

$$\int_0^{\tau} (\psi \cdot \psi)d\tau$$

is a minimum.

10. An Interesting Special Case

The first case to be examined is one in which a simple analytical solution can be obtained. By comparing the iterative solution with the true solution, some insight can be gained into the nature of the technique.

Consider a rocket moving down the tube without rotating. This simple forward motion requires that $\phi = \theta = 0$ for all time. From the definition of β, β is also equal to zero. The equations of motion (7.1a - 7.1f) are reduced to the following simple set

$$(10.1) \qquad \xi' = \varepsilon A \; , \qquad A' = -\nu \; .$$

The solution of Eq. (10.1) can be written immediately as

$$(10.2) \qquad \xi = a_1 + \varepsilon a_0 s - \frac{1}{2}\varepsilon s^2 \nu \; , \qquad A = a_0 - \nu s \; ,$$

where

$$a_1 = \xi(0) \; , \qquad a_0 = A(0) \; .$$

In dimensional terms, the forward motion of the rocket through the smoothbore is

$$(10.3) \qquad x(t) = r_1 a_1 + a_0 \sqrt{\varepsilon r_1 g}\; t - \frac{1}{2}\nu g t^2 \; .$$

For the particular numerical example, let

r_0 = 2.277 inches

r_1 = 2.250 inches

g = 386.2 in./sec^2

ν = 0.06798

and for initial conditions

a_1 = 0.00000

$a_0 = 1.29765$

With this information, the solution was calculated at 0.00(.01).1 and these values were taken as the "observed data" over the time interval 0.0 to 0.11 sec. The results after only one iteration are shown in Table I. While other iterations were calculated, the values did not change after the first.

TABLE I. Linear forward motion

Time (sec)	(Inches)		
	Original Approximation	1st Iteration	Data
0.01	0.202400D-1	0.397863D-1	0.405990D-1
0.02	0.404800D-1	0.780308D-1	0.775900D-1
0.03	0.607200D-1	0.113601	0.113930
0.04	0.809600D-1	0.146492	0.146640
0.05	0.101200	0.176719	0.176750
0.06	0.121440	0.204266	0.204270
0.07	0.141680	0.229139	0.229050
0.08	0.161920	0.251338	0.251280
0.09	0.182160	0.270863	0.270870
0.10	0.202400	0.287713	0.287830

The convergence of the initial conditions and the system parameter, ν, are shown in Table II.

TABLE II. Convergence of the initial conditions and the system parameter

	Original Approximation	1st Iteration	True Value
x(0)	0.000000	-0.113243D-1	0.000000
A(0)	1.276000	1.85577	1.852776
ν	0.000000	0.683893D-1	0.6798D-1

The results of this simple case reveal some interesting points that should be kept in mind when a more complex situation is analyzed. The fact that convergence is attained after only one iteration is not surprising, for the basic system is linear. In comparing the first iteration solution with the "observed data" the fit is only fairly good. Since a minimum least square error fit over the entire region is desired, the error would be expected to be fairly equally distributed over the entire time interval. This does not seem to be true, since an examination of Table I shows that the error is much greater at the zero end of the interval. A plausible reason for this may be shown in Table II where of the three "initial conditions" which must be correctly approximated, $x(0)$ has the greatest error, thus accounting for the distribution of error. This is reasonable when one considers that as the length of the interval over which the data is observed increases, the effect of the initial conditions, $x(0)$, diminishes. On the other hand, as the interval length decreases, the estimate of $x(0)$ should improve but the estimate of ν might be worse for ν is generally determined by the long term behavior of the data. Problems of this kind suggest that the data fit could be improved by using a weighted criterion as discussed in Chapter XV.

BIBLIOGRAPHY AND COMMENTS

1. Section 1. The following book and papers will be useful:

   i. W.J. Harrington and R.E. Bullock, "The Motion of a Spinner Rocket Inside a Smoothbore Launcher", J. Franklin Inst., 277, 552 (1964).

   ii. R. Bellman, H. Kagiwada and R. Kalaba, "Quasilinearization System Identification, and Prediction", RM-3812-PR, (The Rand Corporation, August 1963).

   iii. R. Bellman and R. Kalaba, "Quasilinearization and Nonlinear Boundary Value Problems", (American Elsevier, New York, 1965).

   iv. R. Bellman, B. Gluss and R.S. Roth, "Segmental Differential Approximation and the Black Box Problem", RM-4269-PR, (The Rand Corporation, October 1964).

v. R.S. Roth, "The Unscrambling of Data — Studies in Segmental Differential Approximation", J. Math. Anal. Appl., 14, 5 (1966).

vi. R. Bellman, Dynamic Programming, (Princeton Univ. Press, Princeton, New Jersey, 1957).

vii. R. Bellman and S. Dreyfus, Applied Dynamic Programming, (Princeton Univ. Press, Princeton, New Jersey, 1962).

viii. R. Kalaba, "On Nonlinear Differential Equations, the Maximum Operation, and Monotone Convergence", J. Math. Mech., 8, 519 (1959).

ix. R.E. Bellman, R.E. Kalaba and J. Lockett, Numerical Inversion of the Laplace Transform, (American Elsevier, New York, 1966).

x. R. Bellman, B. Gluss, and R.S. Roth, "On the Identification of Systems and the Unscrambling of Data from some Problems Suggested by Neurophysiology", Proc. Nat. Acad. Sci., 52, 1239 (Nov. 1964).

xi. R. Bellman and R.S. Roth, "Segmental Differential Approximation and Biological Systems: An Analysis of a Metabolic Process", J. Theor. Bio., 11, 168-176 (1966).

xii. E.S. Lee, Quasilinearization and Invariant Embedding, (Academic Press, New York, 1968).

xiii. R.E. Bellman, J. Jacquez, R. Kalaba, and S. Schwimmer, "Quasilinearization and the Estimating of Chemical Rate Constants from Raw Kinetic Data", Math. Biosci., 1, 71-76 (1967).

xiv. R. S. Roth, "Data Unscrambling and the Motion of a Rocket through a Smoothbore Launcher", J. Math. Anal. and Appl., 27, No. 2, 656 (Aug. 1969).

2. Section 2. The technique of differential approximation is described in the book by R. Bellman, Methods of Nonlinear Analysis, Vol. I (Academic Press Inc., New York, 1969).

SUBJECT INDEX

NAME INDEX